The Present and Future of Indoor Navigation

For a listing of recent titles in the
Artech House GNSS Technology and Applications Library
turn to the back of this book.

The Present and Future of Indoor Navigation

Laura Ruotsalainen

Martti Kirkko-Jaakkola

Jukka Talvitie

ARTECH HOUSE

BOSTON | LONDON

artechhouse.com

Library of Congress Cataloging-in-Publication Data
A catalog record for this book is available from the U.S. Library of Congress.

British Library Cataloguing in Publication Data
A catalogue record for this book is available from the British Library.

Cover design by Joi Garron

ISBN 13: 978-1-63081-967-5

685 Canton Street
Norwood, MA 02062

10 9 8 7 6 5 4 3 2 1

Contents

1

Introduction

1.1 Overview

Digitalization is a major technological transformation ongoing in many sectors including transportation, logistics, location-based services, and personal mobility. Global Navigation Satellite Systems (GNSSs) have made outdoor navigation and related services ubiquitous. The ease of navigation has increased the safety of society, when, for example, rescue personnel can find the scene of an accident more easily and quickly with the help of positioning technology. It has also contributed to improving people's quality of life, as when a route in an unfamiliar environment can be found more quickly and stress-free. At the same time, society's infrastructures are becoming more complex: car parking spaces are moving to large underground parking garages and the sizes of public buildings are increasing. Due to the safety and comfort offered by location information, positioning should also work accurately and reliably indoors. In addition, reliable location information obtained indoors would benefit social media applications and companies' business operations, for example, by enabling targeted advertising.

The market research company Maximize Market Research has estimated that the indoor positioning (determining the location of one moment in time on a relevant coordinate frame) and navigation (including the temporal domain and possibly velocity and attitude information in addition to the position) market had a value of 5.98 billion U.S. dollars in 2021 and is expected to reach 102.43 billion by 2029, when other business related to positioning is included [1]. The amount of applications requiring well-performing indoor navigation solutions is vast. Many public areas provide services to help customers and travelers navigate at venues such as shopping malls, as seen in Figure 1.1(a), and hotels, airports,

(a) (b) (c)

Figure 1.1 Applications requiring indoor navigation are vast and exciting: (a) a pedestrian navigating in a mall using a smartphone and two robots that make everyday life more pleasant, (b) a robot vacuum cleaner, and (c) a flying camera broadcasting a sports event.

university buildings, offices, and hospitals. Various location-based services, such as advertisements, discount coupons, and visiting recommendations require indoor navigation. The emergence of cloud-based Internet of Things (IoT) services requires not only accurate indoor navigation solutions, but also seamless transition and therefore continuous navigation from outdoors to indoors. Because the positioning technologies used outdoors and indoors are different, a seamless navigation system requires adaptability to changing environments.

Rescue personnel, police, and tactical operations need accurate and reliable localization and information about a possibly unknown environment. The navigation system in this case must be independent from preinstalled systems as the operations might be required in an unknown and unbounded number of buildings. In many cases the operations require the system to be functional immediately when entering the indoor area and to be tolerant to harsh and rapidly changing conditions of the surrounding space, such as fire and the resulting collapse of structures. In addition to humans navigating themselves, surveillance and monitoring applications in such spaces as mines and public events requires knowledge of other humans' positions and motions. In these cases, we talk about tracking. One very topical example where tracking of humans has been useful is the mapping of the spread of COVID–19 during the pandemic. Tracking applications draw our attention to one significant challenge in the development of navigation systems, securing the privacy of the person being tracked. Although not the topic of this book, privacy protection is essential to successfully entering the navigation market.

Various robots, including autonomous vehicles and drones, increasingly need indoor navigation. In Figure 1.1 there are two robots that make everyday life more pleasant—a robot vacuum cleaner and a flying camera in a balloon broadcasting a sports event. Autonomous vehicles transport people in parking garages and other indoor premises, robots are used for search and rescue, assessment, and reconnaissance in disaster areas, and other tasks that would be risky or impossible for a human to perform. Sometimes the environment sets challenges for both humans and technologies to navigate as seen in Figure 1.2. In addition to humans and robots, asset tracking requires well-performing indoor navigation solution. The retail sector is predicted to have the greatest (annual) growth potential with the adoption of location-based technology and asset tracking [1]. Also, healthcare is estimated to have the fastest growing market share due to increasing adoption of navigation systems for impaired people and their integration with electronic healthcare. Manufacturing, for example tracking goods on production pipelines and logistics, requires highly accurate indoor navigation.

Navigation based on GNSSs is often degraded or completely inaccessible indoors. Although many good indoor navigation technologies already exist, in the future the demand for accuracy, availability, and reliability, among other key performance indicators, will increase due to the emergence of autonomous systems.

While GNSS provides a global and easily accessible navigation solution for most outdoor applications without the need for local infrastructure or calibration, indoor navigation technologies are diverse and their performance is dependent on the application and system setup. Different indoor navigation applications also have different requirements; some examples are listed in Table 1.1. When

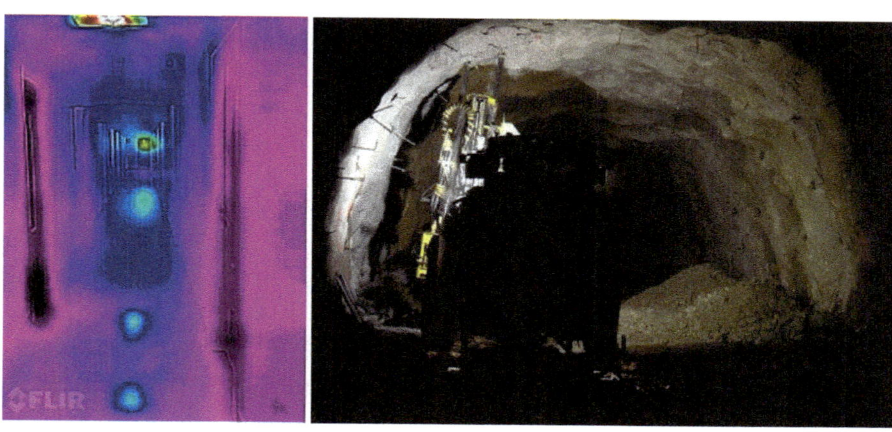

(a) (b)

Figure 1.2 Navigation environments set challenges on the technology: (a) a dark corridor seen via a thermal camera, and (b) a vehicle in an underground mine with a rough surface and low lighting.

Table 1.1
Examples of Indoor Navigation Applications

Use Case	Example Users	Special Needs
Asset tracking	Industry Hospitals Retail	Low unit cost
Personnel tracking	Tactical/rescue operations Mining (safety of life) Retail (customer analysis)	Availability Integrity Minimal user disturbance
Internet of Things	Industry Home automation	Power consumption
Sports analysis	Sports team/coach Referee assistance	Update rate
Robotics	Automated warehouse Robot vacuum cleaner	Environment perception

the requirement for the solution is continuous and the accuracy and reliability requirements are not very strict, setting up a navigation system based on radio signals requiring infrastructure and preparation would be feasible. In contrast, time- and safety-critical rescue operations, especially in unknown environments, prevent relying on equipment deployed in the area. Therefore, navigation must be implemented using sensors carried by the users. There are multiple sensors with different characteristics providing measurements for navigation, such as accelerometers, gyroscopes, barometers, magnetometers, and cameras. At least as important as the selection of the correct technology providing the measurement data for navigation is the development of the algorithm computing the navigation solution. Future autonomous systems will require accuracy and reliability that will set completely new demands for the computation of the navigation solution.

1.2 Preliminaries

We will start the book by introducing certain concepts that appear throughout as they are essential for most indoor navigation setups. First, measurements, data, and tools for obtaining a navigation solution are discussed, then metrics used for assessing the solution performance, concepts of absolute and relative positioning, essential coordinate frames, and finally basic statistics required for calculating the solution.

We have tried to standardize the notation for variables throughout the book. In some cases, the use of corresponding variables in subject areas is so established that we have had to use the same variable in two meanings. Due to the way the subject is treated, there should be no danger of confusion, but the reader should keep this in mind. Such examples are variable K meaning either camera calibration matrix or Kalman gain, depending on the topic, and I for inertial frame or image.

The reader is advised to turn to the following sources when additional knowledge is needed:

- For statistics: [2];
- Navigation and positioning principles: [3];
- Wireless navigation: [4];
- Cellular network architecture: [5];
- Machine learning: [6];
- Deep learning: [7].

1.2.1 Fundamental Means of Indoor Positioning: Measurements, Data, and Tools

Alongside digitalization, the field of indoor positioning has experienced a large range of emerging technological advancements and related new opportunities. For example, through an increased number of varied range of devices that are capable of collecting diverse measurements and sensor data, there is growing potential for considerable improvements in positioning performance. Related to this, recent trends in the signal processing field along with increasing data processing capabilities have led to new positioning solutions and challenged certain conventional approaches. Considering these factors, along with rapid development in potentially available positioning infrastructure, it is possible to achieve improved positioning performance, but only with careful design of the methods used together with use case specific consideration. Overall, it is important to emphasize that as different use cases have different system characteristics and performance requirements, as discussed in the previous section, there is no single positioning solution that could meet the requirements of all use cases.

Modern indoor positioning systems can be built on a wide range of measurements and sensor data. In principle, any measurement that provides information on physical location—uniquely or ambiguously, absolutely or relatively—can be used for positioning. Such measurements and data can be obtained, for example, through radio systems, magnetometers, optical sensors, inertial sensors, barometers, acoustic systems, and even olfaction (smell-sensing) sensors. The actual potential and feasibility of different measurements for a particular use case can vary considerably according to existing infrastructure, possible restrictions in practical implementation, and required positioning performance. For example, radio signals can penetrate through obstacles and propagate behind corners, whereas optical sensors are limited to line-of-sight conditions. On the other hand, compared to radio signals, optical sensors, including light detection and ranging (Lidar) and cameras, are able to provide considerably higher resolution data and support methods like computer vision. Nonetheless, it should be emphasized that different types of measurements or data should not be only seen as alternative or competing, but complementary. To this end, powerful data fusion methods can be considered one of the key components in modern positioning systems.

Due to the increasingly large variety of available measurements and data as well as more diverse positioning system requirements, there is a whole gamut of methods and tools to achieve desired performance objectives. Depending on the use case, employed methods can vary from regression analysis to statistical filtering and machine learning, including various combinations of all. Hence, working with modern and future indoor positioning systems requires increasingly more interdisciplinary consideration and mastering of a multitude of different signal processing techniques.

Conventional regression analysis, including classical least squares methods, is utilized, for example, in radio-based positioning using trilateration and triangulation through range and angle measurements. In addition, regression analysis can be found inside numerous more sophisticated positioning solutions as part of measurement modeling and parameter estimation methods. Both linear and nonlinear regression methods are used, although in many cases nonlinear systems are often linearized for more attractive numerical representation.

The position of a single person or device is highly correlated over time, and therefore it makes sense to estimate consecutive positions as a time series. In other words, assuming a particular scenario-specific statistical characteristics for user movement, the position estimate at certain time step is known to provide information on the following time step. The conventional approach to time-series estimation is statistical filtering, including different variations of Kalman filters, particle filters, and factor graph methods. From these, the most suitable filtering method for a given positioning task can be selected, for example, depending on model complexity, system linearity, and restrictions in computational cost. In addition, statistical filtering is considered an essential tool in sensor fusion, for example, integrating inertial sensor data with measurements from an external positioning system. For instance, dead reckoning, simultaneous localization and mapping (SLAM), and visual odometry are concrete examples of potential spatial filtering applications. Recently, besides statistical filtering, machine learning based methods have become more and more popular in time-series estimation in the context of indoor positioning.

Training-based methods have already been an essential part of indoor positioning systems for a long time. One the most traditional examples of training-based methods is fingerprinting, for example, utilizing measurements of wireless networks and magnetometers. In fingerprinting, the fundamental idea is to collect measurements with location labels from a certain area and then in positioning phase to utilize observed measurements together with the precollected data to estimate the user position. This method is essentially a classification problem, which has been conventionally solved, for example, by a k-nearest neighbors algorithm. However, in recent years, the relatively simple classification task of fingerprinting has been extended to a diverse branch of machine learning technologies, covering more and more complicated scenarios with improved data scalability. Besides fingerprinting, machine learning has been adopted in numerous indoor positioning related solutions due to its built-in ability to handle complicated problems with a large set of unknown and possibly interconnected parameters. In addition, due to its ability to handle interconnected parameters, machine learning has been gradually more exploited also in data fusion.

Although the above discussion on available measurements and potential tools provides insight into potential indoor positioning solutions at an algorithmic level, there are lots of related system-level aspects that affect the system design and need specific consideration in practical implementations. For example, certain positioning solutions require specific infrastructure, which can be either dedicated positioning infrastructure or part of more general infrastructure, including communications networks. Moreover, certain positioning solutions allow user devices to operate stand-alone without necessary interactions with other devices. On the other hand, some solutions require centralized data processing, and therefore means of communications between devices. Overall, there are several system-level questions that dictate the feasibility of positioning solution for a given use case: who obtains the measurements or collects data, who estimates the position, where the actual numerical computations are carried, and who possesses the required auxiliary information for position estimation. Depending on the answers, the performance of the positioning system changes, for example, in terms of privacy, security, required signaling (or feedback) overhead, latency, and power efficiency.

1.2.2 Navigation Performance Metrics

The most important parameters defining the performance of an indoor navigation system are availability, accuracy, and continuity. Availability is the percentage of time that the services of the system are usable, accuracy is the closeness of an estimate to its true value, and continuity is the share the system functions without interruption of the total navigation time. With the emergence of safety-critical indoor navigation applications, there is also the fourth main performance metric from the GNSS domain, integrity, which is the measure of the trust that can be placed in the correctness of the information supplied by a navigation system, which is increasing in importance. As a GNSS-domain example, the vertical accuracy requirement for precision aircraft landing could be 8 meters at 95% confidence, but in terms of integrity, the navigation system must ensure that the probability of a gross altitude measurement error (e.g., 50m) must remain below a set threshold. If the available set of measurements is inadequate to meet the integrity requirement, the system must raise a timely integrity alarm to the user. A method for integrity monitoring was developed for an indoor navigation system integrating ultrawideband (UWB) and inertial navigation system (INS) measurements [8]. The method, called innovation-based integrity monitoring (IBIM), uses innovations or residuals calculated as a difference between each UWB time difference of arrival (TDoA) measurement and predicted position obtained using INS measurements. Measurements are rejected if the difference between the calculated residual vector and the lowest residual threshold is higher than the confidence interval.

Different kinds of errors affect the quality of obtained measurements. The errors can be divided into two main groups: stochastic errors and biases. The effect of stochastic errors can be expressed in terms of accuracy and precision. Accuracy demonstrates the closeness of an estimate to its true value and precision the closeness of observations to their mean. In Figure 1.3, the true value is the center of the dartboard (i.e., the bullseye), and the black dots are the obtained measurements. Absolute accuracy is defined as the accuracy of a position with respect to a well-defined reference coordinate system. In some indoor navigation scenarios, it is enough to determine the position with respect to, for example, the initial navigation point without fixing that to a coordinate system and in such cases we talk about relative accuracy.

In addition to accuracy, availability, continuity, and integrity, other relevant metrics must be taken into account. Acceptable navigation system installation time and cost are dependent on the application. Some systems require

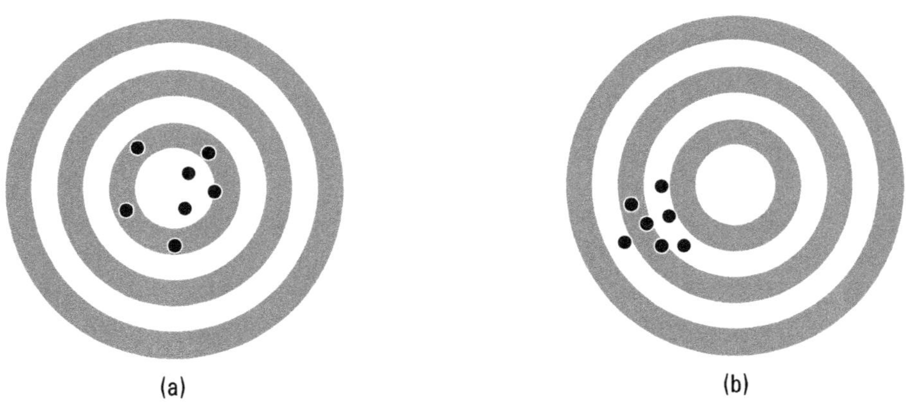

Figure 1.3 Accuracy and precision of a navigation system: (a) shows a result that is accurate, but not precise, namely the measurements (black dots) are close to the true value (bullseye) but not to each other, and (b) the result is precise but not accurate.

maintenance or recalibration. For example, the performance of certain navigation systems based on radio signals may be disturbed by changes in the surroundings and thereby require tuning. Energy consumption is related to the sustainability of the system as well as the maintenance burden. For example, heavily power-consuming radio frequency (RF) identification (RFID) tags used for tracking of people and assets might need to be changed frequently to keep the system running. Latency, namely the time between sending a location request and receiving the location information, is relevant for time-critical navigation systems [3, 9].

1.2.3 Absolute and Relative Positioning

The position solution computed using different technologies may be divided into two classes: absolute and relative, and their computation methods into position fixing and dead reckoning, respectively [3]. Position fixing means that the system determines the position directly relative to a set coordinate frame. Dead reckoning in turn uses a set initial position and propagates it with measurements sensing the motion of the user. The measurements used in computation may be either *external information* such as signals transmitted with different radio technologies such as Wi-Fi or Bluetooth, or *environmental features*, such as optical information, magnetic fields, or air pressure.

At present, indoor navigation technologies using radio signals are the most widespread. The reason is mainly their availability due to their utilization in other relevant functions. Such signals are called signals-of-opportunities (SoOps), referring to signals that are initially meant for other purposes, such as Wi-Fi for wireless communications, but can also be used for positioning. SoOps provide roughly meter-level accuracy with respect to a predefined coordinate frame. However, systems based on SoOps require installation before use as well as continuous calibration.

Dead reckoning based navigation is challenging because incremental position propagation is susceptible to error accumulation. At present, there is no single solution that would provide required accuracy and reliability such as for indoor navigation at rescue operations. Time- and safety-critical operations often take place in unknown environments, which prevents relying on equipment placed into the area such as Wi-Fi transmitters. Therefore, navigation must be implemented using sensors carried by the users. In many cases, the users are pedestrians, and therefore the equipment must be lightweight and low-cost. Such requirements create fundamental challenges on the long-term navigation performance. The application area requires real-time functioning with mobile devices, which excludes several computation methods. Infrastructure-free navigation, namely using a system that is able to localize itself independent of any equipment preinstalled to the building, builds upon using various sensors monitoring the motion of the user. Fusion of the sensor measurements for propagating an initial position solution provides continuous relative positioning [10, 11]. Reliable and accurate infrastructure-free indoor navigation is still an unsolved issue when the indoor time is not tightly bounded. Cost and size requirements of pedestrian navigation system demands use of microelectromechanical system (MEMS) sensors, which results in measurement errors and rapidly drifting position solutions. Therefore, effort has been put into developing data fusion algorithms for mitigating the effect of errors and extending the eligible navigation time indoors.

Figure 1.4 presents different indoor navigation technologies addressed in this book with respect to their accuracy, availability, and system implementation cost. Selecting the most suitable technology is dependent on the requirements set by the application, environment, and other case-specific constraints.

1.2.3.1 Accuracy Requirements for Absolute and Relative Position and Navigation

As absolute and relative navigation solutions are based on very different principles, their accuracy requirements must also be defined differently. The common absolute accuracy requirement for nominal consumer applications is at the meter level. However, the requirement set for autonomous vehicles is much more strict: outdoors at the 20-cm level and most likely will be indoors also at the level of decimeters. Many absolute and relative navigation applications tolerate less accuracy, and often the requirement is set to room level or for vertical solutions at floor level. For example, for tracking assets at hospitals, room information might be sufficient to locate the device. As will be discussed throughout the book, relative navigation systems suffer from drift, meaning that

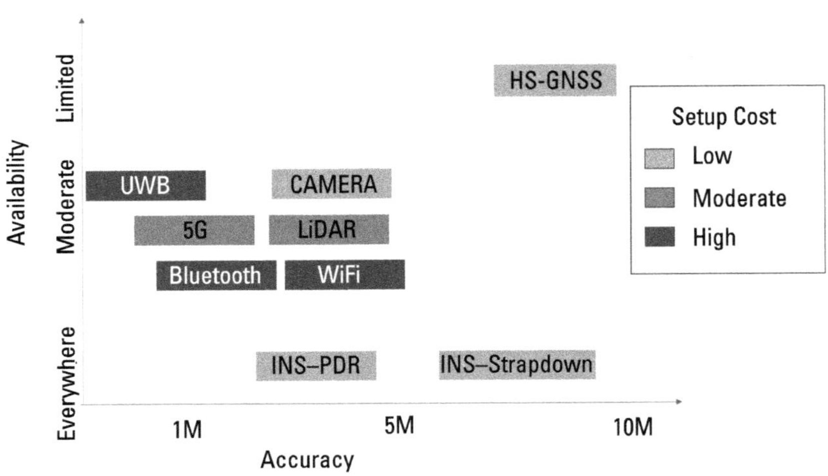

Figure 1.4 Indoor navigation systems and their accuracy, availability, and system implementation cost.

the measurement and computation errors accumulate in time and therefore the absolute accuracy boundaries are not relevant. However, for a relative solution the most used accuracy measure is the percentage error. At the time of writing this book the state-of-the-art relative indoor navigation methods obtain the level of 1% error.

1.2.4 Coordinate Frames

In order to express positions, the first thing one needs is a coordinate frame. At a global scale, it is common to use the geographic latitude–longitude–altitude system. While this coordinate system is easily human-understandable, it is often inconvenient for positioning algorithms as it is polar: for instance, it is not trivial to calculate the distance between two sets of latitude and longitude. Cartesian (rectangular) coordinate frames have certain advantages in this sense as one can, for instance, use Euclidian norms as target functions for optimization algorithms. A global Cartesian coordinate frame can be defined by first setting the origin at the center of mass of the Earth, and directing the x-axis toward the zero longitude meridian in the equatorial plane. The z-axis points toward the North Pole along the axis of rotation, and finally, the y-axis is defined such that it complements the right-handed rectangular coordinate system; the result is illustrated in Figure 1.5(a). This coordinate frame is called the Earth frame (E-frame); its main drawback is that the coordinates are not very human-intuitive: for instance, it takes a bit of calculation to even know if a point is above or below the Earth's surface.

For indoor navigation, global coordinates are usually not so significant: it is more important to know the position with respect to some frame fixed to the building. In this respect, a Cartesian coordinate frame with origin at a relevant location in the building and leveled x- and y-axes is very convenient. We can define the local level frame (L-frame) by pointing the x-axis toward East, the y-axis to North, and the z-axis upward; this coordinate frame is commonly referred to as the East–North–Up (ENU) frame and is shown in Figure 1.5(b). In some applications, the positioning system does not know where the North is, making

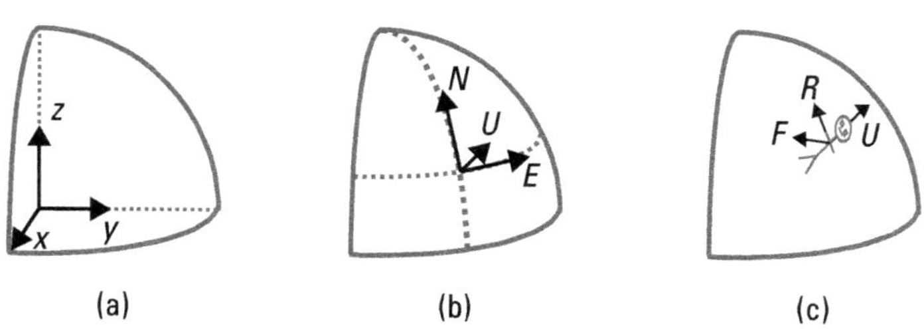

Figure 1.5 Illustration of the most important coordinate frames: (a) E-frame, (b) L-frame, and (c) B-frame (right-forward-up).

it impossible to use the ENU system. In such cases, the direction of the x-axis at the beginning of operation can be taken as the azimuth reference.

Many sensors produce information that is relative to a coordinate frame defined by the sensor itself, which is called the body frame (B-frame). For instance, an inertial navigation system commonly uses three accelerometers mounted perpendicular to each other, and a camera perceives information within its field of view. In practice, the B-frame is often fixed to the user, for example, in terms of the direction of travel such as right–forward–up (Figure 1.5(c)). In order to use the sensor measurements, they need to be converted to the navigation frame; this procedure involves translation and rotation. Between Cartesian coordinate systems, the rotations can be applied by multiplying the three-dimensional (3D) coordinate vectors with rotation matrices, also known as *direction cosine matrices*. A rotation matrix R is defined as a 3×3 orthogonal matrix (i.e., $\mathrm{R}^T\mathrm{R} = \mathrm{R}\mathrm{R}^T = \mathrm{I}$) with $\det \mathrm{R} = +1$. As such, rotation matrices can also be used to express orientations: they describe the offset of the coordinate axes with respect to a reference orientation (e.g., ENU axes).

The drawback of using rotation matrices for orientation representation is poor human readability. This problem is commonly mitigated by factoring the matrix to a sequence of elementary rotations with respect to some reference axes. For instance, if we define the B-frame of a pedestrian with x pointing to the right, y forward, and z up, then the elementary rotations along these axes are known as *pitch*, *roll*, and *heading* (or *yaw*)—commonly referred to as *Euler angles*. Then, the rotation matrix R_B^L, referring to the rotation from the B-frame to the L-frame, would be factored for pitch p, roll r, and heading ψ as

$$\mathrm{R}_B^L(p, r, \psi) = \begin{bmatrix} \cos\psi & -\sin\psi & 0 \\ \sin\psi & \cos\psi & 0 \\ 0 & 0 & 1 \end{bmatrix} \begin{bmatrix} \cos r & 0 & -\sin r \\ 0 & 1 & 0 \\ \sin r & 0 & \cos r \end{bmatrix} \begin{bmatrix} 1 & 0 & 0 \\ 0 & \cos p & -\sin p \\ 0 & \sin p & \cos p \end{bmatrix}. \tag{1.1}$$

This formulation implies that each successive rotation in the sequence is relative to the coordinate axes where the earlier rotations have been applied. Therefore, the order of the rotations is important in the Euler angle representation and these rotations are sometimes referred to as *intrinsic*.

It is noteworthy that the Euler angle representation is not a one-to-one mapping of rotation matrices. First, different intrinsic rotation sequences exist that yield the same rotation matrix. Moreover, if the second intrinsic rotation $r = \pm 90°$, the third rotation will be applied effectively along the same axis as the first, implying the loss of one degree of freedom in the Euler angle representation. This is not necessarily a problem in pedestrian applications where the roll and pitch angles usually remain small, but the phenomenon is inherent to Euler angles. In fact, a complete representation of orientation requires a minimum of four parameters; unit quaternions are an example [12]. However, in the remainder of this book, we will use rotation matrices to represent orientations and rotations unless the use of Euler angles is justified in a particular case.

In the context of inertial sensors, yet another coordinate frame is necessary to be defined. Inertial sensors measure acceleration and rotation with respect to an *inertial frame* (I-frame) which is a nonrotating, nonaccelerating coordinate frame. It is important to note that since we only measure motion with respect to the I-frame, with the measurement expressed in the sensor B-frame, there is no need to define the coordinates of the origin or the orientation of the reference axes of the I-frame.

1.2.5 Basic Statistics

Many positioning algorithms are based on statistical characterization of the measurement uncertainty. In this section, we briefly present some basic concepts of statistics that occur in the most common positioning systems.

Suppose $\boldsymbol{x}$ is a random vector with mean value $\boldsymbol{\mu}$ and covariance matrix Σ. Then, given a constant matrix A and vector $\boldsymbol{b}$, respectively, the random variable $\boldsymbol{y} = \mathrm{A}\boldsymbol{x} + \boldsymbol{b}$ has mean value $\mathrm{A}\boldsymbol{\mu} + \boldsymbol{b}$ and covariance $\mathrm{A}\Sigma\mathrm{A}^T$. This identity is widely used for error propagation through functions of random variables $\boldsymbol{y} = f(\boldsymbol{x})$. Often the functions are nonlinear, in which case the coefficient matrix A is approximated by the Jacobian of the function f.

Models of measurements $\boldsymbol{y}$ are often defined as probability distributions conditional on the position (or another parameter of interest) $\boldsymbol{x}$, denoted as $P(y|x) = \frac{P(\boldsymbol{x} \text{ and } \boldsymbol{y})}{P(\boldsymbol{x})}$ where the distribution of the measurement uncertainty is assumed known. From this definition we can derive the well-known Bayes' law for the conditional probability of $\boldsymbol{x}$:

$$\begin{aligned} P(x \text{ and } y) &= P(y|x)P(x) = P(x|y)P(y) \\ P(x|y) &= \frac{P(y|x)P(x)}{P(y)} \end{aligned} \tag{1.2}$$

By applying Bayes' law, we can combine *prior* information of the unknowns $\boldsymbol{x}$ with measurements $\boldsymbol{y}$ to update our information about the distribution of $\boldsymbol{x}$. For instance, when estimating the position of the user at consecutive epochs of time, the position at the previous time instant can be converted to prior information about the current location, given assumptions (i.e., a model) of the motion. In turn, the previous position estimate is conditional on the previous measurements; this is the basic principle of recursive (or sequential) Bayesian filtering.

It is a widespread practice to use multivariate normal (Gaussian) distributions to model uncertainties in positioning algorithms. A normal distribution is parameterized by a mean value $\boldsymbol{\mu} \in \mathbb{R}^n$ and a covariance matrix $\Sigma \in \mathbb{R}^{n \times n}$, the probability density function being

$$P(\boldsymbol{x}) = \frac{1}{\sqrt{(2\pi)^n \det \Sigma}} \exp \frac{-(\boldsymbol{x}-\boldsymbol{\mu})^T \Sigma^{-1} (\boldsymbol{x}-\boldsymbol{\mu})}{2}. \tag{1.3}$$

A random variable $\boldsymbol{x}$ following a normal distribution with mean $\boldsymbol{\mu}$ and covariance Σ is denoted by $\boldsymbol{x} \sim \mathcal{N}(\boldsymbol{\mu}, \Sigma)$.

Although the normal distribution is not an exact characterization of all uncertainties related to positioning systems, it has certain advantages that facilitate its use in algorithms: for instance, the sum of two normally distributed random variables is also normally distributed. This makes it possible to propagate the uncertainties as normal distributions, and recalling that only the mean value and covariance matrix are needed to characterize the distribution, the computations can usually be conducted relatively efficiently. Moreover, the *central limit theorem* states that the average of any N independent and identically distributed random variables approaches a normal distribution as N tends toward infinity; this justifies the applicability of normal distributions as long as there are a sufficient number of measurements.

Unfortunately, normal distributions also have characteristics that cause challenges with certain models. First, the distribution is always symmetric about the mean value; this does not fit well to, for example, radio signal time of arrival (TOA) measurements where anomalies such as reflections can only *increase* the propagation time. Second, the probability density of a normal distribution is quite concentrated around the mean, leaving little probability to the tails of the distribution. This can compromise the robustness of the positioning algorithm: if the measurement uncertainty is modeled as Gaussian, occasional gross measurement faults can corrupt the position estimate because the measurement model considers it unlikely that a measurement would be so strongly erroneous. Third, normal distributions are always unimodal; this is a problem in scenarios where there are not sufficient measurements to determine a unique solution.

Figure 1.6 compares the standard normal distribution $\mathcal{N}(0, 1)$ with some alternatives: a lognormal distribution defined as $\log x \sim \mathcal{N}(0, 1)$, a standard Cauchy distribution characterized by $P(x) = \left(\pi + \pi x^2\right)^{-1}$, and a Gaussian mixture distribution that is a weighted superposition of several normal distributions. We can see that each of these examples addresses one of the shortcomings of the normal distribution defined above, but adopting them would also increase the algorithmic complexity; this is the reason why normal distributions are usually preferred if possible.

When statistic models are used, an important concept is the *residual*, which is the difference between the realized value of a measured quantity and the value

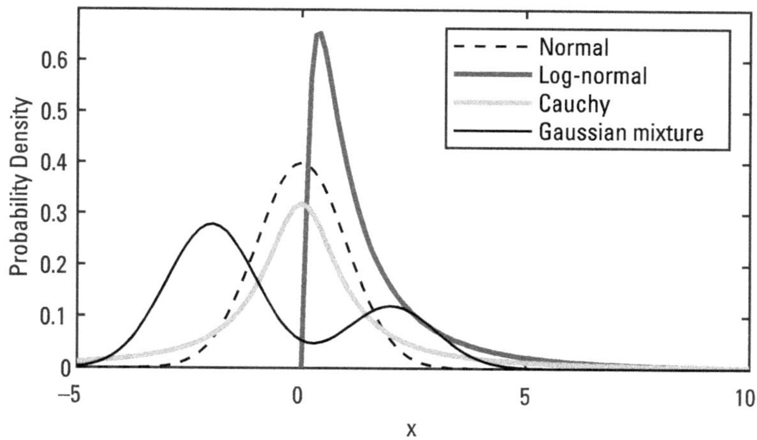

Figure 1.6 Comparison of the standard normal distribution with nonsymmetric, heavy-tailed, and multimodal distributions.

predicted by the model. Many estimation algorithms, such as least squares, are based on minimizing residuals. Furthermore, statistical tests can be conducted on the residual itself to detect measurements that do not conform to the model (i.e., outliers): for instance, a set of measurements that includes outliers is mutually inconsistent, which often leads to large residuals.

1.2.6 Contents of the Book

Chapter 2 will present the different systems providing positioning measurements, external information, and environmental features. Chapter 3 will discuss how such measurements may be further processed to be used in forming the navigation solution, both absolute and relative. Finally, Chapter 4 will describe how to set up different navigation systems. This book's focus is on present technologies and algorithms, but the material within also provides a look into future possibilities. Each chapter provides the essential mathematics as well as practical examples to support both the theoretical understanding and ability to set up an actual indoor navigation system.

References

[1] Maximize Market Research, *Indoor Positioning and Indoor Navigation Market–Global Industry Analysis and Forecast (2022–2029)*, www.maximizemarketresearch.com.

[2] Wasserman, L., *All of Statistics: A Concise Course in Statistical Inference*, New York: Springer, 2010.

[3] Groves, P. D., *Principles of GNSS, Inertial, and Multisensor Integrated Navigation Systems*, Second Edition, Norwood, MA: Artech House, 2013.

[4] Bensky, A, *Wireless Positioning Technologies and Applications*, Norwood, MA: Artech House, 2016.

[5] Garcia, A. C., S. Maier, and A. Phillips, *Location-Based Services in Cellular Networks: From GSM to 5G NR*, Second Edition, Norwood, MA: Artech House, 2020.

[6] James, G., D. Witten, T. Hastie, and R. Tibshirani, *An Introduction to Statistical Learning: With Applications in R*, New York: Springer, 2014.

[7] Goodfellow, I., Y. Bengio, and A. Courville, *Deep Learning*, Cambridge, MA: MIT Press, 2016, http://www.deeplearningbook.org.

[8] Ascher, C., L. Zwirello, T. Zwick, and G. Trommer, "Integrity Monitoring for UWB/INS Tightly Coupled Pedestrian Indoor Scenarios," in *2011 International Conference on Indoor Positioning and Indoor Navigation*, IEEE, 2011, pp. 1–6.

[9] Van Haute, T., E. De Poorter, P. Crombez, et al., "Performance Analysis of Multiple Indoor Positioning Systems in a Healthcare Environment," *International Journal of Health Geographics*, Vol. 15, No. 7, 2016.

[10] Ruotsalainen, L., M. Kirkko-Jaakkola, J. Rantanen, and M. Makela, "Error Modelling for Multi-Sensor Measurements in Infrastructure-Free Indoor Navigation," *Sensors*, Vol. 18, No. 2, 2018.

[11] Ruotsalainen, L., A. Morrison, M. Makela, J. Rantanen, and N. Sokolova, "Improving Computer Vision-Based Perception for Collaborative Indoor Navigation," *IEEE Sensors Journal*, Vol. 22, No. 6, 2022, pp. 4816–4826.

[12] Kuipers, J. B., *Quaternions and Rotation Sequences: A Primer with Applications to Orbits, Aerospace and Virtual Reality*, Princeton, NJ: Princeton University Press, 2002.

2

Positioning Measurements, Sensors, and Their Errors

This chapter presents the different systems providing positioning measurements that can further be used in forming a navigation solution. These systems can roughly be divided into two classes: those needing setup of the infrastructure before being used for navigation, and those based only on equipment carried by the user and providing ad hoc information. The former class consists mainly of different transmitters producing radio signals, and receivers acquiring those and computing the navigation solution via various methods. In navigation, one main characterization point for signals is the frequency. The requirements set for a radio signal based navigation system as well as its performance vary greatly depending the signal and its frequency used. The frequency bands the signals propagate at span from Global Navigation Satellite System (GNSS) 1.57 GHz to 5G's 40 GHz. Indoors radio signal propagation suffers from multipath, fading, reflections, and deep shadowing effects. The higher the frequency, the larger the multipath and penetration loss effects, but the antenna can be made physically smaller.

In Section 2.1, we will discuss various radio positioning systems available for indoor navigation, some providing better and some worse accuracy. These radio positioning systems are Global System for Mobile Communications (GSM), Universal Mobile Telecommunications System (UMTS), Long Term Evolution (LTE), 5G New Radio (5G-NR), Wi-Fi and Bluetooth, UWB, and high-sensitivity GNSS. Radio signals usually provide an absolute navigation solution in a previously set-up navigation frame. The second class providing relative ad hoc information contains different sensors providing information about the user's or information's state or its change. Navigation using sensors usually requires estimation of an initial position and orientation that is further propagated with the received measurements. Section 2.2 will discuss the most important sensors used in indoor navigation, which are inertial sensors, magnetometers, barometers, and optical systems. As the methods turning the measurements from optical systems into motion information are quite complex, we will dedicate Section 2.3 to discussing the scientific field encompassing the methods that comprise computer vision.

2.1 Radio Signals

When available, RF signals are highly useful for positioning applications. GNSS is an example of a radio navigation system that was originally developed for positioning. In fact, many different radio signals can be leveraged for positioning even if they were actually developed for a completely different purpose, in which case they are called signals of opportunity. For instance, detecting and identifying a transmitter gives proximity information, and the received signal strength correlates with the distance. Wireless communication infrastructure tends to stay in the same place for a long time and transmit permanent identifiers such as medium access control (MAC) addresses that can be turned to position information if the transmitter locations are known. Furthermore, many modern wireless communication systems support features and signals that are designed for positioning purposes (e.g., round-trip time (RTT) or angle of arrival (AOA) measurements), which further improves the positioning performance that can be achieved.

In the following sections, we first present the evolution of cellular networks and the positioning features they can offer. Cellular networks (or mobile networks) consist of a core network for overall network management and a radio access network for enabling radio communications capability between the network and user devices. From these we focus on the radio access network, as it provides the physical means for positioning signals and measurements. The main idea of cellular networks is to divide a geographical area of the network into cells to efficiently distribute the available radio resources, such as frequencies, between different devices. Then, during active communication phase, each user device is associated with a specific serving cell operated by a specific base station (BS). Because of user mobility, cellular networks introduce a mechanism for handovers that aims at providing a continuous communications experience when moving from one cell to another. To orchestrate such a complicated network, there are numerous reference signals available in cellular networks that are suitable for positioning purposes. Moreover, from 4G networks onward there have been dedicated signals for positioning purposes available. One of the great benefits of using cellular networks for positioning is their extensive coverage, including both outdoors and indoors. After cellular networks, we discuss short-range radio signals, including Wi-Fi, Bluetooth, and UWB. Finally, the challenges involved in the use of GNSS indoors are outlined.

2.1.1 GSM

The digital era of mobile networks started when the deployment of GSM networks accelerated at the beginning of the 1990s. The main motivation for the new GSM networks was to replace old analog voice telephony technologies with a new global standard. Because of this, the GSM was originally operating solely based on circuit-switched connections but was later updated to also support packet-switched data along with the introduction of General Packet Radio Service (GPRS). With circuit-switched connections, communications are performed over a dedicated communications channel, which needs to be established separately before each communication session. However, modern communications systems are based on packet-switched connections for more efficient utilization of a communication channel.

The GSM, standardized by the European Telecommunications Standards Institute (ETSI), operates at different frequency bands, including 850-MHz, 900-MHz, 1,800 MHz, and 1,900 MHz bands [1, 2]. The lower bands, namely the 800 MHz and 850 MHz include a total of 50-MHz bandwidth, which is equally split into downlink (DL) and uplink (UL) bands according to the frequency division duplex (FDD) principle. In mobile networks, DL refers to transmission from a BS to a mobile station (MS), whereas UL refers to transmission from a MS to a BS. Furthermore, the upper 1,800-MHz and 1,900-MHz bands have a total of 75 MHz for the DL and 60 MHz in UL. The maximum transmission power of the MS varies between 0–30 dBm for the MS and 34–58 dBm for the BS depending on the used band. For all the bands, the MS and BS reception sensitivities are defined as −104 and −102 dBm, respectively.

For communications purposes each MS is associated with a specific DL and UL carrier from the available GSM frequency band, where different carriers are separated by 200 kHz. In radio communications systems, the term carrier refers to a waveform that carries the information-bearing signal, such as bits in digital communications. The same carrier can be used by multiple MSs based on a time division multiple access (TDMA) scheme. This means that MSs that have simultaneous radio communication channels with the network are separated based on the time of transmissions. In one TDMA frame of 4.6-ms duration, there are eight time slots, in which the actual BS and MS transmissions can occur in the form of bursts. A burst is transmitted using Gaussian minimum shift keying (GMSK) modulation with a symbol rate of 1625/6 kHz $\approx$ 270.8 kHz.

From the positioning perspective the GSM introduces four separate positioning methods referred to as [3]:

- Enhanced observed time difference (EOTD);
- Uplink time of arrival (UTOA);

- Timing advance (TA);
- Assisted GNSS (A-GNSS).

From these the EOTD and UTOA methods are based on measuring a one-way radio propagation delay between the BS and MS, whereas the TA method is based on a measurement of a round-trip time. Whereas the EOTD, UTOA, and TA methods are stand-alone GSM positioning methods, the A-GNSS method requires a GNSS receiver in the MS to exploit GNSS measurements together with additional network-assisted data.

EOTD is a DL-based method that exploits observed time difference (OTD) taken at the MS and real time difference (RTD) measurements at the network side location measurement units (LMUs). The OTD measurements indicate the time interval between the reception of training sequences from bursts of 2 BSs. Besides positioning purposes, the OTD is used in GSM for communications-related tasks, such as for synchronization during handovers from one BS to another. If the BSs are synchronized, the OTD can be directly used for positioning based on the time difference of arrival (TDOA) principle, as illustrated in Figure 2.1. By using the TDOA method, it is possible to neglect the clock offset between the MS and the network by observing propagation time differences between different BSs. As a result, the position estimate can be found at the intersection of hyperbolas as shown in Figure 2.1. If the BSs are not synchronized, the network, more specifically the LMU, needs to provide the MS an additional RTD measurement that reveals the clock difference between the BSs related to a particular OTD

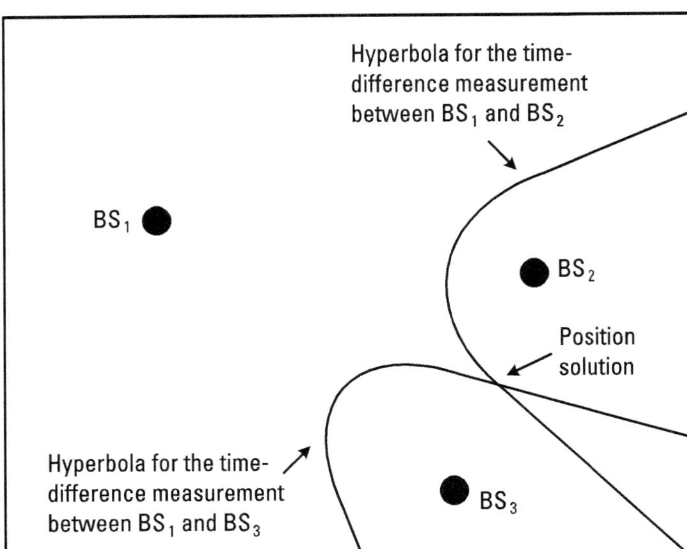

Figure 2.1 Illustration of the TDOA-based positioning.

measurement. The numerical accuracy of both the OTD and RTD measurements are equal to 1/256 of the symbol duration, which is approximately 14.4 ns. This reflects roughly to a distance resolution of 4.3m. However, in practice the accuracy is limited due to relatively small transmission bandwidth, multipath propagation, and often low signal power levels. Related to this, it should be noted that in cellular networks obtaining positioning measurements from multiple BSs typically means that the majority of the measured BSs signals are received outside of the intended cell coverage areas.

UTOA could be considered an UL-based counterpart of the EOTD, where the position estimate is attained at the network side. In UTOA, TOA measurements are collected at the network side based on UL bursts of the MS. A clear difference compared to the EOTD framework is that the measurements at the network should be obtained by the LMUs, whereas in EOTD, OTD measurements are extracted from regular BS transmission. As a result, UTOA requires the presence of more LMUs in the network architecture. Nonetheless, similar to EOTD, UTOA exploits the TDOA principle to obtain the MS position estimate, and thus removes the requirement for a time-synchronized MS.

TA-based positioning utilizes the TA measurement whose sole existence originates from a fundamental communications requirement related to support multiple MSs in the TDMA frame. Since MSs are generally located at different distances from the BS, the MSs perceive different DL frame timings due to different radio propagation delay. As a result, UL transmissions of each MS need to be uniquely aligned according to the TA so that they are received inside the desired time slot at the BS. In practice, MSs with larger distances to the BS begin their burst transmissions earlier (with respect to their own time slot) than the MSs closer to the BS. TA is fundamentally a RTT measurement, and measures the range between the MS and BS. Thus, the position estimate can be obtained by using the multilateration principle, as shown in Figure 2.2. In multilateration, the position estimate is found at the intersection of BS-centered circles that represent the measured ranges between the between the MS and each BS. The actual range measurements can be obtained by using various techniques, for example, TA, RTT, or TOA. Regarding the TA-based positioning, it should be emphasized that TA measurement was not originally designed for positioning purposes, and thus lacks accuracy by providing only 3.69 μs timing accuracy. This relates to a distance resolution of about 550m. However, one benefit of the TA is that it is continuously measured and available when the MS is active, such as, during an ongoing voice or data call.

Besides the above-described stand-alone GSM positioning methods, GSM includes a non-stand-alone A-GNSS method, where the mobile network is used to improve the GNSS positioning performance. From the GNSS perspective, BS equipment is often installed at advantageous locations with higher altitudes compared to typical MS operating altitudes. This enables use of GNSS receivers

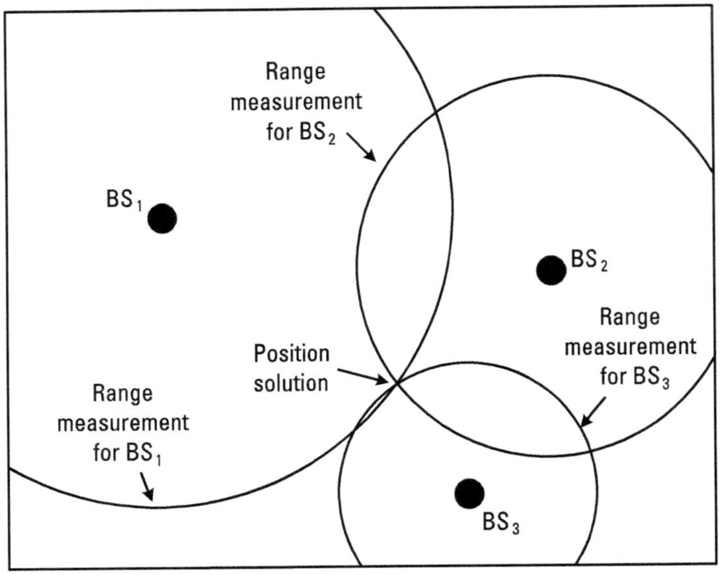

Figure 2.2 Illustration of the multilateration-based positioning that can be used with various types of range measurements, such as TA, RTT, and TOA/time of flight (TOF).

at known BS locations, where there is a clear satellite visibility and subsequent ability to continuously monitor the status of various GNSS parameters. With A-GNSS, the GSM network can provide the MS-based GNSS receiver with specific assistance data to reduce the GNSS time-to-first-fix, and improve the GNSS accuracy by different correction parameters. Due to generally poor indoor coverage of GNSSs, A-GNSS can be especially useful for indoors. Utilizing GNSS indoors is discussed in more detail in Section 2.1.8.

On top of the positioning-related measurements utilized by the standardized GSM positioning method, there are MS-based RXLEV measurements that relate to the distance between the MS and BS via pathloss models [4–7]. The RXLEV measurement indicates the received signal power (dBm) of the BS broadcast control channel measured at the MS. Compared to other measurements, the benefit of RXLEV measurements is that they are continuously monitored and available in both idle and active MS operation. This is because the RXLEV measurements have a crucial role in both connected and idle state mobility management procedures. Furthermore, RXLEV measurements are often available via an application programming interface of different mobile device platforms, which makes the RXLEV a desirable option for third-party positioning services. However, similar to all other received signal strength measurements, RXLEV is sensitive to signal blockage and various dynamic channel fluctuation in the radio propagation environment. In addition to pathloss-based positioning approaches, RXLEV measurements are often considered in fingerprinting-based systems [8–10].

2.1.2 UMTS

3G UMTS networks, specified by the 3rd Generation Partnership Project (3GPP), answered to the increasing need of early mobile broadband in the late 1990s. The new UMTS technology introduced a new radio access network with a new BS type, called Node B, and a new MS type, called user equipment (UE). However, principally the same core network architecture was maintained in the first UMTS release. Compared to GSM, one of the most significant differences in the UMTS radio interface is the multiple access technology used, which in UMTS is based on Code Division Multiple Access (CDMA). Compared to the TDMA used in GSM, in CDMA different UEs having a simultaneous network access are separated by specific spreading codes. This means that DL signals from different Node Bs, or correspondingly UL signals from different UEs, are received at the same time in the same frequency. Furthermore, besides the familiar FDD method from the GSM, UMTS introduced an option for a Time Division Duplex (TDD) method for separating the DL and UL transmissions in time.

Together with the introduction of UMTS standard [11–14], new frequency bands were specified about in the same frequency range with the GSM, the highest carrier frequency locating at 2,100 MHz. For the data transmission, UMTS uses phase shift keying (PSK) or quadrature amplitude modulation (QAM) modulation with a chip rate of 3.84 MHz and carrier separation of 5 MHz. The duration of one UMTS FDD frame in DL is 10 ms, which results in a total of 38,400 chips/frame. Compared to GSM, the dynamic range of the UE maximum transmission power is a bit higher, 21–33 dBm, which is especially useful to handle some of the crucial power control mechanism in CDMA. Similarly, there is a wider scale defined for the Node B transmission powers, ranging from 20 dBm to unlimited power depending on a considered cell coverage category of Node B. The required receiver sensitivities for Node B and UE are between −107 and −121 dBm, and −114 and −117 dBm, respectively.

UMTS standard [15] describes several positioning methods, including:

- Cell ID and enhanced cell-ID;
- Observed time difference of arrival (OTDOA);
- Uplink-time difference of arrival (U-TDOA);
- A-GNSS.

In addition to these, positioning methods utilizing non-UMTS-based measurements are specified. These include use of barometric pressure sensors and other wireless technologies, including wireless local area network (WLAN), Bluetooth, and a specific positioning dedicated terrestrial beacon system. When available, these measurements can be used together with the UMTS-measurements, or individually, to improve the position performance.

The cell ID method is based on simply detecting the identity of a cell, where the UE is currently located. In cellular networks, depending on the UE state, the UE location might be known in a level of one cell or multiple cells. The former condition occurs typically when the UE has an active voice or data call, whereas the latter case is typical when the UE is in an idle state without any active connection to the network. During an idle state, the UE periodically monitors a specific paging channel, which is used by the network to reach an idle UE. Now, since in the cell-ID based positioning the identity of the serving cell is needed, a paging procedure is required for an idle state UE to reveal the cell ID. Alternatively, it is also possible to determine the UE location roughly based on a specific service area identifier, which covers a geographical area of multiple cells. A particular feature of UMTS is a soft handover, where the connection of active UE to the network operates through branches of multiple cell IDs. Thus, for the soft handovers the cell ID used for the location determination can be selected based on various criteria, such as signal quality level, signal power level, RTT measurements, or for example, the order of setting up different cells for a soft handover.

The positioning accuracy of the cell ID method is proportional to the area of a cell, whose radius can vary from hundreds of meters to tens of kilometers. Enhanced cell-ID methods introduce additional measurements that can be used together with the cell ID to improve the positioning accuracy. At the UE side, these measurements include pathloss for the FDD and TDD modes, and UE Rx-Tx time difference, common pilot channel (CPICH) received signal code power (RSCP) and CPICH energy per chip over the noise spectral density (E_c/N_0) for the FDD mode. Moreover, at the Node B side, the possible measurements are the RTT in the FDD mode, and Rx timing deviation and AOA in the TDD mode. Here, the Rx-Tx time difference determines the time difference between the beginning of UE dedicated physical control channel (DPCCH) UL transmission and the first detected path of the received downlink-dedicated physical channel (DPCH). Furthermore, the CPICH RSCP is a measurement of the received signal power from Node B, similar to RXLEV in GSM. The CPICH RSCP and pathloss measurements can be used for estimating the distance between Node B and the UE according to pathloss models. The required accuracy of CPICH RSCP is between ±6–8 dB. Based on the CPICH RSCP and the measured total received power, it is possible to derive CPICH E_c/N_0, which is a measurement of signal quality. The RTT measurement is performed at the network side and it indicates the difference in time between the transmission of the DPCH frame in downlink, and the reception of the corresponding DPCCH frame in UL. Together with the

UE Rx-Tx time difference, the RTT enables estimation of the two-way radio propagation delay. The required accuracy of the RTT measurements is half of the chip duration, but it can be reported in resolution of 1/16 of chip duration. This enables distance resolution of about 5m. When using the RTT to estimate the physical distance between Node B and the UE, the measurement of UE processing time (UE Rx-Tx Time Difference) is especially crucial due to the assumption of clock drift at low-cost UEs. In the TDD mode, the enhanced cell-ID method can utilize the Rx timing deviation measurement at Node B. The Rx timing deviation measures the time between the first received path of UE UL transmission and the time of the beginning of the respective transmission slot according to internal timing of Node B. Besides the Rx timing deviation, the TDD mode supports AOA measurements, which define the estimated angle of the UE with respect to a given reference direction, standardized as North by the 3GPP. In addition to the above-described standardized measurements, the enhanced cell-ID method can utilize different RF pattern-matching based techniques using various types of RF measurements, including the ones described earlier.

Although different variations of the enhanced cell-ID method can exploit various positioning related measurements from ranging to angle estimation, it is fundamentally designed for obtaining measurements from a single serving cell. However, certain variants of the enhanced cell ID are able to utilize CPICH RSCP from neighboring cells or multiple serving cell measurements during a soft handover. Nevertheless, to obtain accurate positioning-related measurements from multiple cells, UMTS standard introduces the OTDOA positioning method, which can be directly used for TDOA-based positioning. In OTDOA, the UE obtains type 2 system frame number (SFN) observed time difference measurements, which indicate the time difference between the beginning of reception of CPICH signal from two separate cells. When the CPICH RSCP is larger than −94 dBm, the accuracy of the type 2 SFN-SFN observed time difference is half of the chip duration, which refers to a ranging resolution of 40m. However, because of the CDMA principle used in UMTS, it is difficult to accurately measure transmission from multiple Node Bs simultaneously. For example, when the UE is close to the serving cell, simultaneous neighboring cell measurements at the same frequency can be easily blocked by the serving cell transmissions. Therefore, in order to enable gathering of measurements from multiple cells, the UMTS includes an option for so-called idle period DL, where each Node B stops its transmission for short periods of time. During these periods, the UEs can measure nonidle Node Bs without the severe interference from the other Node Bs. Similar to the EOTD in GSM, for asynchronous Node Bs it is also required to measure the RTD, which can also be potentially performed during the idle periods. The OTDOA method can be used in either the UE-based or UE-assisted approach, where in the former the UE performs the positioning calculation, and in the latter the UE performs the measurements and reports them back to the network.

The U-TDOA and network-assisted GNSS are essentially based on the same approaches as in the GSM, and therefore they are not discussed here. UMTS-specified positioning measurements related to WLAN and Bluetooth are discussed in more detail in Sections 2.1.5 and 2.1.6, respectively. Similar to GSM, some of the UMTS measurements, such as the CPICH RSCP, are often available via the application-programmable interface of different platforms. This makes UMTS a potential option for signals-of-opportunity-based positioning as well as also for third-party positioning service providers, for example, via fingerprinting-based approaches.

2.1.3 LTE

To answer to the ever-increasing demand for high data rate mobile broadband services, the 4G LTE standard was introduced in the late 2000s. The 3GPP-specified 4G evolution also included LTE-Advanced and LTE-Advanced Pro, which introduced additional performance improvements [16]. LTE technology presented a completely new network architecture comprising both the core network and the radio access network. The familiar circuit-switched connections from earlier mobile network generations were omitted, and the new LTE network was built purely on packet-switched internet protocol-based network architecture. In the LTE radio access network, the base station is referred to as evolved Node B (eNB) and the user device as UE.

The physical (PHY) layer of the LTE radio interface is based on the Orthogonal Frequency-Division Multiplexing (OFDM) principle using QAM-modulated subcarriers [17]. This approach also works as the foundation for the Orthogonal Frequency-Division Multiple Access (OFDMA) scheme used in the LTE DL. Similar to 3GPP, LTE enables both FDD and TDD for duplexing, but unlike in UMTS, all positioning-related measurements are available in both duplexing methods. There are numerous available frequency bands standardized for the LTE, and all the frequency bands are within the so-called frequency range 1 (FR1) region, defined between frequencies 410–7125 MHz [18, 19]. One of the main differences in LTE compared to earlier mobile network generations is that LTE supports a variable transmission bandwidth, from 1.4 MHz to 20 MHz, which allows network operators to use scattered spectrum resources more efficiently. The length of the LTE frame is 10 ms, which is identical to the one used in UMTS. Each frame is divided into 10 subframes, and each subframe includes 2 slots. Furthermore, the sampling frequency is a multiple of the chip rate in UMTS and is defined as 30.72 MHz. The subcarrier spacing is fixed to 15 kHz and the cyclic prefix of the OFDM symbol is given as 160 samples for the first symbol in the slot and 144 samples for the remaining symbols in the slot. The maximum number of subcarriers for the 20-MHz bandwidth allocation is defined as 1,200, which results in an effective bandwidth of 18 MHz. For the single carrier transmissions, the maximum UE transmission power varies from 23 to 26 dBm depending on the given UE power class [19]. Moreover, the maximum eNB transmission powers varies from 14 dBm to unlimited power depending on the number of transmit antenna ports and on the considered base station coverage profiles, including wide area, local area, and home base stations [18]. It should be noted that in the LTE standard many of the design parameters, such as the maximum transmission power and receiver sensitivity, vary according to different device configurations, for example, related to carrier aggregation and multiple-input and multiple-output (MIMO).

LTE introduces several standardized positioning methods, such as downlink positioning, enhanced cell ID method, and uplink positioning, which all utilize LTE-originated measurements [20]. Furthermore, the A-GNSS, WLAN, Bluetooth, terrestrial beacon system, and sensor-based methods exploit supporting measurements from other systems. These non-LTE-based methods are very similar to the ones already specified for UMTS and are not considered here further.

The downlink positioning method is based on OTDOA, which utilizes reference signal time difference (RSTD) measurements between different cells. The RSTD measurements are taken at the UE side based on DL transmitted positioning reference signals (PRSs). The PRS is the first ever dedicated positioning-related reference signal introduced in the 3GPP-specified mobile networks. The PRSs can be configured for transmissions with different periodicity, and can be transmitted during the naturally existing idle periods of the other LTE reference signal and control channel transmissions. The standardized accuracy requirement for the RSTD measurements is dependent on the used bandwidth and vary between $\pm 4T_s$ and $\pm 15T_s$, where T_s is the LTE time unit $T_s = 1/(15000 \times 2048)$ seconds. However, the measurement reporting supports a resolution of 1 time unit, which roughly maps the maximum ranging accuracy of the downlink positioning method to 10m.

The enhanced cell ID method of the LTE is fundamentally the same with the UMTS enhanced cell ID method and can use various measurements to improve plain cell ID based positioning. In LTE it is possible to utilize the new LTE-based measurements, such as reference signal received power (RSRP), reference signal received quality (RSRQ), received signal strength indicator (RSSI), AOA, and TA. From these the RSRP indicates the received signal strength of a cell at the UE based on DL cell-specific reference signal (CRS) transmission. In normal conditions the required measurement accuracy of the RSRP is specified between 4.5–8 dB using a reporting range from −156 to −44 dBm with 1-dBm resolution. On the other hand, the RSSI is a measurement of total received power, which includes the cumulative received power from other cell transmissions considered as interference. The reporting range for the RSSI measurement is given from −100 to −25 dBm with 1-dBm resolution. Based on the RSRP and RSSI, RSRQ can be calculated to reflect the signal quality. Besides the signal power measurements, the enhanced cell ID method of LTE can utilize TA measurements, which are needed to achieve UL synchronization during the random access procedure. Similar to the TDMA frame timing in GSM, TA measurements can be used to synchronize the reception of UL signals, originating from UEs at different distances to the eNB receiver, to the correct slots in the LTE frame. The TA measurement defines the total round-trip delay including the processing time at the UE. Therefore, together with the UE Rx-Tx time difference measurement, the round-trip radio propagation delay can be solved. The reporting resolution of the TA-based round-trip propagation delay is $2T_s$, which equals to one-way distance resolution of roughly 10m.

The LTE uplink positioning method is based on the uplink relative time of arrival (UL-RTOA) measurements performed by the LMUs at the network side, which is very similar to UTOA and U-TDOA used in GSM and UMTS, respectively. In LTE, the measurements are obtained based on sounding reference signals (SRSs) transmitted by the UE in UL. The required UL-RTOA measurement accuracy differs according to the used bandwidth, the used carrier aggregation configuration, and the number of multiple parallel UE transmissions. Affected by these the specified accuracy requirement for the UL-RTOA varies between 6–10 T_s which is defined as a 90% percentile error. However, the UL-RTOA reporting allows the accuracy of up to 2 T_s which results in spatial resolution of 20m.

2.1.4 5G NR

The 5G New Radio (NR) is a new 3GPP-specified mobile network standard that aims at providing services to a wide variety of use cases and different industry verticals. Besides high data rates, other performance indicators, such as latency, reliability, and scalability are an essential part of the 5G NR design. From the beginning, to enhance the deployment of new 5G NR networks, 5G NR has been specified to operate fluently with the LTE standard. In fact, the first 5G NR specifications were only including operation in a non-stand-alone mode that required presence and management by the LTE network. The first stand-alone 5G NR was introduced with release 15 in 2018 [21]. The 5G includes a comparable radio network architecture with the LTE. In the 5G NR context, the user device is still called UE, but the BS, instead, is referred to as next generation Node B (gNB). However, it should be emphasized that in the 5G NR era, the physical network device performing the RF processing and the device running the related software are not necessarily the same due to the introduction of a centralized radio access network and cloud-based computing [22]. Therefore, it is important to consider the gNBs as logical network elements.

A fluent interaction between the LTE and 5G NR is supported with the similarity of many PHY layer parameters. Most importantly, similar to the LTE DL, the 5G NR PHY layer is built on the OFDM-based transmissions [23]. Moreover, the fundamental frame structure in 5G NR shares the same main parameters with the LTE, including a frame duration of 10 ms and subframe duration of 1 ms. Like LTE, the 5G NR supports both TDD and FDD modes, from which the TDD approach has gained lots of attention due to its benefits, for example, regarding MIMO transmissions. Besides the FR1 bands familiar to the earlier mobile network generations, 5G NR introduces new frequency bands from Frequency Range 2 (FR2), containing frequencies from 24.25 to 52.6 GHz [24]. For the 5G NR, one of the clear benefits of the FR2 is the availability of large bandwidths that are specified up to 400 MHz, whereas the maximum bandwidth at the FR1 is 100 MHz. Since 5G NR is required to support various use cases, as well as higher carrier frequencies, the 5G NR radio interface has more flexible design compared to LTE. This can be seen, for example, in variable subcarrier spacing, flexible bandwidth allocations, and related reference signal structures. The subcarrier spacing in 5G NR can be configured as 15, 30, 60, 120, or 240 kHz, which directly affects the OFDM symbol duration and the number of OFDM symbols in the subframe. In general, compared to LTE, in 5G NR the role of MIMO techniques has increased, and MIMO processing is a central part of many basic 5G NR network functionalities, including the transmission of cell synchronization signals with synchronization signal blocks (SSBs) [23].

Positioning has been in an important role during the whole development of 5G NR specifications. 5G NR introduces a relatively wide set of standardized positioning methods [25]:

- A-GNSS methods;
- OTDOA positioning based on LTE signals;
- Enhanced cell ID methods based on LTE signals;
- WLAN positioning;
- Bluetooth positioning;
- Terrestrial beacon system (TBS) positioning;
- Sensor-based methods (barometric pressure sensor and motion sensor);
- NR enhanced cell ID methods based on NR signals;
- Multicell round-trip time (multi-RTT) based on NR signals;
- Downlink angle of departure (DL-AOD) based on NR signals;
- Downlink time difference of arrival (DL-TDOA) based on NR signals;

- UL-TDOA based on NR signals;
- Uplink angle of arrival (UL-AOA), including azimuth-AOA and zenith-AOA based on NR signals.

From the above methods, the last six are based on 5G NR signals, while the others have been passed down from LTE to 5G NR.

The NR enhanced cell ID methods are very similar to the one in LTE, and the main difference lies in incorporating some new 5G NR related measurements. Especially regarding the received signal strength measurements, there is a significant difference in 5G NR compared to LTE. Whereas in LTE, the RSRP measurement can be used to estimate the range between the gNB and UE via pathloss models, in 5G NR the corresponding range estimate has built-in sector information. This is because in 5G NR the received signal strength measurement; that is, the synchronization signal (SS)-RSRP, is measured from a single SSB transmission that is beamformed to a specific direction. In normal conditions, the measurement accuracy of the SS-RSRP varies between ±4.5 and ±8 dB [26]. The SS-RSRP reporting can be performed in layer 1 and layer 3. Regarding layer 3, the reporting range spreads from −156 to −31 dBm with 1-dB resolution.

Multi-RTT positioning is a new 5G NR positioning method, where the positioning is based on obtaining timing measurements at both the UE side and network side. For multi-RTT positioning, the measurements are obtained based on the PRSs in DL and SRS signals in UL. The corresponding measurements include at least the UE Rx-Tx time difference and gNB Rx-Tx time difference measurements, and optionally DL-PRS-RSRP and UL-SRS-RSRP measurements. The required accuracy of the UE/gNB Rx-Tx time difference measurements are not yet specified, but the reporting enables resolution of $4T_c$ in FR1 and T_c in FR2, where T_c is the basic 5G NR time unit defined as $T_c = 1/(480000 \times 4096) \approx 0.509$ ns [26]. Thus, the delay estimates in multi-RTT positioning are able to provide a ranging resolution of about 60 cm in FR1 and 15 cm in FR2. DL-PRS-RSRP and UL-SRS-RSRP measurements define the received signal strength of the DL PRS and UL SRS. The detailed specification of DL-PRS-RSRP accuracy is still ongoing, but at least the UL-SRS-RSRP can be reported with the same range than the previously described SS-RSRP (i.e., −156 dBm...−31 dBm with 1-dB resolution).

The angle-based positioning methods in 5G NR include the DL-AOD and UL-AOA methods, which are both based on the spatial processing at the gNB side antenna. Based on angle measurements, the positioning can be performed based on the angulation principle, as illustrated in Figure 2.3. In the angulation method (also often referred to as triangulation or multiangulation), the position estimate is found at the intersection of the angle-vectors pointing out from each BS toward the measured angle. In 5G NR-based DL-AOD the angle between the UE receiver and gNB transmitter is estimated based on DL-PRS-RSRP

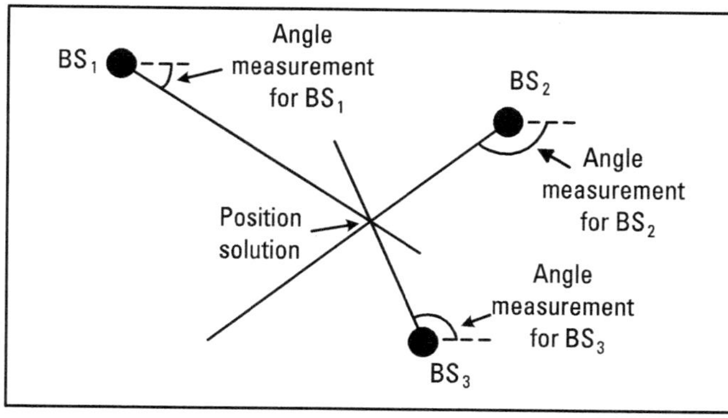

Figure 2.3 Illustration of the angulation-based positioning for AOA and AOD measurements.

measurements taken at the UE from the DL PRS. Together with the spatial knowledge of each DL PRS transmission, the angle of departure (AOD) can be estimated. Since the specification of DL-PRS-RSRP is not yet finished, the measurement accuracy requirements and corresponding reporting ranges of DL-PRS-RSRP are still unknown. The UL-AOA method is based on network side measurements relying on UE transmitted SRS in UL. In this case, the actual angle estimation method is left as an implementation-specific issue, and therefore the related accuracy requirements for estimating the AOA are not provided in the 5G NR specifications. However, the angle estimates can be reported with as high as 0.1-deg resolution covering the full angular domain in the azimuth and elevation directions [26]. Besides the angle estimates, the UL-AOA can also exploit the UL-SRS-RSRP for improving the estimation accuracy.

The DL-TDOA method is based on obtaining time-difference measurements of the received DL PRSs from different gNB transmitters at the UE. In 5G NR, these measurements are referred to as RSTD measurements and they have the same timing report resolution than the UE/gNB Rx-Tx time difference measurements in the multi-RTT method (i.e., $4T_c$ in FR1 and T_c in FR2, resulting in a distance resolution of about 60 and 15 cm, respectively). For positioning, besides the RSTD measurements, it is required to know the relative timing between different network side transmitters. In addition, DL-PRS-RSRP can be used in aiding the timing measurements and the corresponding positioning solution.

UL-TDOA can be considered the UL counterpart of the DL-TDOA. In UL-TDOA, the time differences are estimated based on the UL SRSs transmitted by the UE and received by different gNB receivers. The obtained time differences are reported as UL-RTOA measurements, whose reporting resolution is T_c, corresponding roughly to distance of 15 cm, for both FR1 and FR2. Besides the UL-RTOA measurements, UL-SRS-RSRP measurements can be optionally used to improve the performance.

In 5G NR, the multitude of positioning methods and available reference signals [23] enable countless hybrid positioning approaches, which can adopt aspects on different positioning methods, positioning-related measurements, and/or reference signals from the above-discussed standardized approaches. Due to the large set of enabled positioning approach variations, some of the measurements that can be potentially used for positioning purposes (e.g., TA) have not been discussed here.

Beyond 5G and upcoming 6G, systems are expected to improve the localization performance further [27]. The important technical enablers for the enhanced positioning performance in future networks include, for example, new higher-frequency bands (allowing THz imaging), reflective intelligent surfaces, efficient beam space processing, and machine learning techniques. Besides performing only positioning-related tasks, future positioning systems will be part of integrated positioning and sensing functionalities as well as SLAM settings. Furthermore, aspects of security and privacy are expected to play a more crucial role in future system design.

2.1.5 Wi-Fi

Wi-Fi, more officially characterized by the Institute of Electrical and Electronics Engineers (IEEE) WLAN 802.11 standard, delivers wireless local area connectivity especially targeted at indoors. Unlike the mobile cellular networks operating in licensed frequency bands, Wi-Fi systems are commonly operating at unlicensed bands, from which the most familiar ones are probably the 2.4-GHz industrial, scientific, and medical (ISM) band and the 5-GHz band. Operating in unlicensed bands makes the Wi-Fi signals subject to interference from other systems, and therefore greatly affects different design aspects in the standardization, including positioning-related aspects. Because of the extensive deployment and coverage of Wi-Fi access points, Wi-Fi-based positioning has been one of the most studied indoor positioning methods in the existing literature [28–34]. The comprehensive IEEE WLAN 802.11-2020 standard [35] includes different PHY layer implementations and has been continuously building up through different amendments during the past years. In a stack of radio communication protocols, the PHY layer is the layer closest to a radio antenna, and provides means for transmitting raw bits between radio devices. In general, Wi-Fi-based positioning methods can be generally divided into received signal strength based approaches [28–30] and timing-based approaches [31, 33, 34]. In addition, angle-based positioning methods are also possible when equipped with a MIMO setup [36]. In the Wi-Fi context, the base station is referred to as an access point (AP) and the user device as a station (STA).

Probably the most frequently used measurement in different Wi-Fi positioning methods is the received signal strength measurement. In Wi-Fi, there are multiple received signal strength measurements that indicate slightly different things. Perhaps the most used signal strength measurement is the beacon RSSI, which measures the signal strength of the AP beacon periodically broadcasted by different APs. As specified by the standard, the Wi-Fi RSSI has an accuracy of ±5 dB defined as a 95% confidence interval [35]. If the receiver has multiple MIMO branches, the RSSI is defined as the mean of all branches. Because of the relatively small coverage area of indoor Wi-Fi APs, often solely detecting APs with a known location (similar to cellular-based cell ID methods) can provide relatively good position accuracy. However, by considering a proper pathloss model, the RSSI measurement can be associated with a specific radio propagation distance [5], but signal blockage in the radio path can severely affect the RSSI and induce error in the distance estimate. In addition, RSSI values in devices originating from different vendors might present significant RSSI offsets with respect to each other. For this reason, a dedicated calibration of RSSI measurements might be needed in order to achieve satisfying positioning performance [32].

Besides using the RSSI measurements for pathloss-based positioning, much of the research has been conducted on RSSI-based fingerprinting methods [37]. In fingerprinting, the RSSI measurements are first collected during an offline phase (or training phase), and then possibly processed, and finally stored into a database. After this, during the online phase (or positioning phase), the STA can be positioned based on comparing the current measurements against the previously stored database. Since in practical fingerprinting-based positioning systems one can assume that measurements from both the training phase and positioning phase are collected with numerous different user devices originating from different vendors, an appropriate RSSI calibration method has a central role in proving good positioning performance. During recent years, there has also been a dramatic increase in research applying machine learning methods to the RSSI fingerprinting-like positioning context [38–40]. Although both conventional fingerprinting and machine learning consider the learning phase and positioning phase, machine learning methods do not require extensive databases, but attempt to fit learned data into the considered model.

As part of the IEEE 802.11mc amendment, which is currently incorporated into the 802.11-2020 standard [35], Fine Time Measurement (FTM) was introduced to enable accurate RTT-based positioning with Wi-Fi devices. However, already before this, timing measurements were specified in the 802.11-2012 standard for enabling clock synchronization between Wi-Fi devices. These timing measurements enabled distance estimation between the STA and the serving AP but could not obtain distance estimates to other APs to obtain find a unique position solution.

FTM measurements are based on a two-way ranging (TWR) procedure, as illustrated in Figure 2.4, where the t_1, t_2, t_3, and t_4 define the measured time stamps for computing the RTT. These timing measurements can be reported

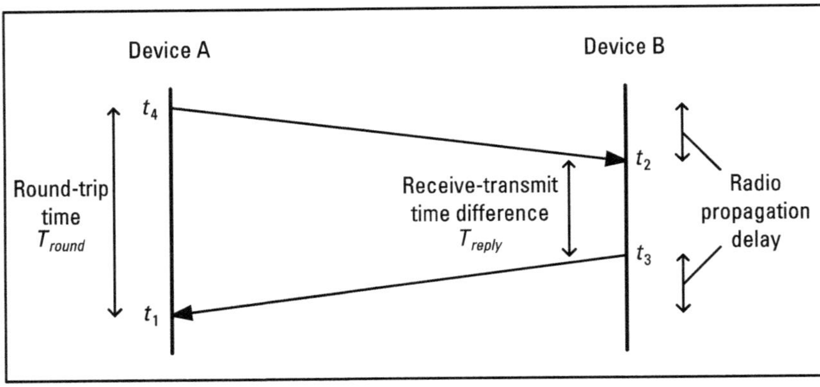

Figure 2.4 Illustration of the TWR procedure.

with as high as 1-ps resolution, which corresponds to 3-cm distance. However, already due to limitations of transmission bandwidth and clock drifts, the practical measurement accuracy is clearly worse [41]. The maximum bandwidth for the FTM measurement is 80 MHz, but in many cases smaller bandwidths, such as 20 or 40 MHz, are utilized.

According to empirical studies [42], there might be a significant bias in the FTM-estimated ranges that need to be removed by a separate calibration method or at least considered in the subsequent positioning solution. Since Wi-Fi devices are often operating indoors, the direct line-of-sight (LOS) path between the devices is often absent because of considerable signal obstruction either caused by a fixed physical object or a moving object or person. This, together with multipath propagation, induces error in the FTM ranging. In many practical scenarios, the FTM-based approaches are reported to have a meter-level ranging accuracy [34, 41, 43].

The next significant improvements for Wi-Fi-based positioning are expected along the upcoming 802.11az amendment [44], often referred to as next generation positioning (NGP). The new NGP approach provides an extended maximum bandwidth of 160 MHz with the possibility of increasing the bandwidth up to 320 MHz by the introduction of future Wi-Fi generations. In addition, NGP can more efficiently utilize spatial domain MIMO measurements via beamforming techniques. Together with the increased bandwidth and enhanced MIMO capability, NGP should be able to provide better performance and increase tolerance against multipath propagation. Furthermore, NGP also enables more efficient ranging of multiple STAs by a single AP transmission, and improves PHY and MAC layer security.

2.1.6 Bluetooth

Bluetooth is a wireless standard for low-range communication using the same 2.4-GHz ISM band as Wi-Fi. The Bluetooth standard [45], prepared by the

Bluetooth Special Interest Group (SIG), can be roughly divided into two main branches: classic Bluetooth (version 2.0) introduced in 2004, and Bluetooth Low Energy (BLE) introduced in 2009. In the following discussion we focus on BLE (versions 4.x and 5.x) due to its supported positioning capabilities. Currently, BLE is found in numerous mobile devices, which makes it an appealing option for several positioning use cases and scenarios. Moreover, the low power consumption and low cost of BLE devices, combined with relatively good positioning accuracy, separates BLE from other wireless positioning technologies.

BLE transmissions occur at the unlicensed 2.4-GHz ISM band, more specifically between frequencies 2.402–2.480 GHz. Inside this 78-MHz frequency interval, there are 40 channels with 2-MHz channel spacing. From the 40 channels in total, 37 channels are for data transmission and 3 channels are for advertising. In BLE, the advertisement channels serve the same types of purposes as the beacons in Wi-Fi. BLE transmission utilize, frequency hopping spread spectrum (FHSS) with Gaussian frequency shift keying (GFSK) modulation. Based on the FHSS principle, BLE transmissions are hopping between different channels in order to increase frequency diversity and consequently to reduce the effect of interference. The raw PHY layer data rates of BLE are specified up to 2 Mbit/s, but with coded transmission the net user data rate is reduced. Moreover, the maximum transmission power of BLE devices is 20 dBm and the minimum required receiver sensitivity is −70 dBm.

The main positioning-related measurements provided by BLE are RSSI, along with AOA and AOD, which are supported in the standard starting from Bluetooth version 5.1. However, detecting the presence of a BLE signal from a known location already enables a relatively accurate position solution due to the typically short radio propagation range of BLE transmissions. Related to this, BLE proximity detection, similar to the cell ID based methods in cellular networks, has been widely studied in literature, especially motivated and driven by the COVID-19 pandemic [46, 47]. The accuracy of cell ID and proximity-based detection methods are greatly dependent on the underlying network density. Nonetheless, in order to improve the accuracy of plain cell ID or proximity-based positioning, the distance between the BLE transmitter and BLE receiver can be estimated based on the RSSI measurements and appropriate pathloss models. The required RSSI measurement accuracy in BLE is specified as ±6 dB. However, there is an important detail in obtaining the RSSI measurements that should be considered when operating with certain BLE devices. In practice RSSI is measured distinctly from the three separate advertisement channels that are separated in the frequency by more than 20 MHz. In many channel conditions, frequency separation of this size results in frequency selective fading between the advertisement channels, and consequently the observed RSSI are different at each band. Therefore, depending on the way the BLE device reports the RSSI, even in static channel conditions, there might be significant fluctuation in the reported RSSI due to channel-wise differences [48, 49].

Bluetooth version 5.1 introduced a method for phased antenna array based AOA and AOD estimation for BLE devices [45]. For both angle estimation methods, the estimation can be performed in a connected state or in a connectionless state. In the connected state, angle estimation is done based on regular link layer transmission, whereas in the connectionless state measurements of the advertising channels are used. Since signal wavelength is one of the essential parameters in the considered phased-array-based angle estimation approach, the signal wavelength should be constant during the signal reception. However, because of the GFSK modulation scheme used in BLE, the transmit frequency changes according to each transmitted bit. Thus, to enable the phased-array-based angle estimation, a specific constant tone extension (CTE) field, comprising only bits of 1, is added in the end part of the transmitted data packet. In this way the frequency (and wavelength) remains constant during the CTE, allowing accurate angle estimation. There are no accuracy requirements for estimation of AOA and AOD in the BLE standard. The accuracy is dependent on the used antenna array sizes and can considerably vary between different positioning scenarios. In the upcoming BLE standardization, high-accuracy distance measurements are also envisioned for further improving BLE positioning performance [50].

2.1.7 Ultrawideband

As stated by the International Telecommunication Union Radiocommunication Sector (ITU-R) in [51], UWB refers to a technology for short-range radio communications typically having the −10-dB bandwidth of at least 500 MHz, or a fractional bandwidth (ratio between the −10-dB bandwidth and carrier frequency) of more than 0.2. The term UWB itself does not directly relate to any specific standard, and several UWB standards have been introduced [52]. However, different standards generally consider different Open Systems Interconnection (OSI) model layers, and as usual, there is no guarantee for interoperability between separate standards. However, common existing UWB standards are based on the same IEEE 802.15.4-2020 standard [53], which defines PHY and MAC layer functionalities for low-rate wireless networks. Since the PHY layer parameters are in an essential role in determining the positioning capability, we focus on the IEEE 802.15.4 standard in the following discussion.

Regarding positioning, UWB technology has the advantage of large bandwidth, which enables accurate timing and ranging estimation as well as providing tolerance against multipath and interference. In IEEE 802.15.4-2020, the UWB PHY layer can operate in two modes: high-rate pulse (HRP) repetition frequency and low-rate pulse (LRP) repetition frequency, where in the HRP mode there is the optional ranging feature described in the standard. In total there are 16 available carrier frequencies for the HRP transmission ranging from 499.2 to 9,484.8 MHz. Furthermore, for each specified carrier frequency, there is an explicitly determined transmission bandwidth varying from 499.2 to 1,354.97 MHz so that the carrier frequency is always a multiple of the bandwidth. The HRP transmissions are carried out with a combination of burst position modulation (BPM) and binary phase-shift keying (BPSK) modulation with a minimum data rate of 0.11 Mbit/s, but possibly reaching up to 27.24 Mbit/s depending on the channel coding scheme used. The main difference of BPM compared to traditional carrier-modulated signals is that in BPM the data bits are mapped into positions of different pulses. The maximum transmission power is not explicitly described, but it is stated that the maximum output power spectral density should be restricted as specified by the appropriate regulatory bodies. The requirement for the receiver sensitivity is defined as −45 dBm/MHz, which delivers smaller than 1% packet error rate (PER) for packets with a length of 20 octets.

An enhanced version of the UWB PHY and MAC layer as well as the associated ranging techniques was introduced in 802.15.4z [54] as an amendment to the main IEEE 802.15.4-2020 standard. The ranging-capable devices (RDEVs) referred to in the IEEE 802.15.4-2020 were updated to RDEVs in 802.15.4z with the possibility to support ranging in both LRP and HRP modes (LRP-enhanced ranging-capable devices (ERDEVs) and HRP-ERDEVs). The associated UWB ranging methods include, one-way ranging, TWR, and TDOA. All the approaches are based on observing the transmit and receive timing of the extremely narrow pulses enabled by the UWB. In the case of one-way ranging, the transmitter and receiver must share the same information of time, which typically requires additional network infrastructure to manage interdevice synchronization. Alternatively, if the mobile device has an unknown time but there are multiple synchronized nodes in the network, the TDOA principle can be used for positioning. Furthermore, the synchronization requirement between two devices can be completely neglected when using TWR techniques, like RTT in 5G NR.

In 802.15.4z [54], multiple separate TWR techniques are considered: single-sided two-way ranging (Ss-TWR), double-sided two-way ranging (Ds-TWR), and Ds-TWR with three messages. The basic principle of the Ss-TWR method is similar to the one used with Wi-Fi FTM, as illustrated in Figure 2.4. The propagation delay (i.e., TOF) between device A and B can be estimated based on the measured round-trip time T_{round} at device A and the measured reply time T_{reply} at device B as

$$\hat{T}_{propagSS} = \frac{1}{2}\left(T_{round} - T_{reply}\right). \tag{2.1}$$

Since T_{reply} and T_{round} are measured at devices A and B separately with unique clock frequency offset errors, the propagation delay estimate $\hat{T}_{propag}$ can have a significant error especially when reply time measurements T_{reply} become large. To decrease the error resulting from the clock frequency offsets between the ranging devices, Ds-TWR is proposed, where two round-trip measurements are obtained in consecutive manner. With Ds-TWR, the estimated propagation delay can be written as

$$\hat{T}_{propagDS} = \frac{T_{roundA} \times T_{roundB} - T_{replyB} \times T_{replyA}}{T_{roundA} + T_{roundB} + T_{replyA} + T_{replyB}}, \quad (2.2)$$

where T_{roundA} and T_{roundB} are the round-trip times, and T_{replyA} and T_{replyB} are the reply times, measured by devices A and B in respective order. Ds-TWR using four messages, as shown in Figure 2.5, can be reduced to Ds-TWR with three messages by using the first reply of device B to initiate the second round-trip time measurement.

The maximum timing resolution of UWB is extremely accurate. For example, with the HRP mode, the ranging measurements can be reported by a specific ranging counter using resolution of 1/128 of the 499.2 MHz chipping period. This is approximately 15.7 ps in time, which reflects to 4.7 mm in distance. However, for low-cost and low-power-consumption devices, employing the required PHY layer counter running at 64 GHz is not often feasible, and thus, in practice the reported ranging counter used above the PHY layer has synthetically lower resolution. In [55], per the IEEE P802.15 Working Group for Wireless Personal Area Networks, typical errors of the propagation delay estimation by using the TWR approach are reported to be at a level of 0.1–4 ns

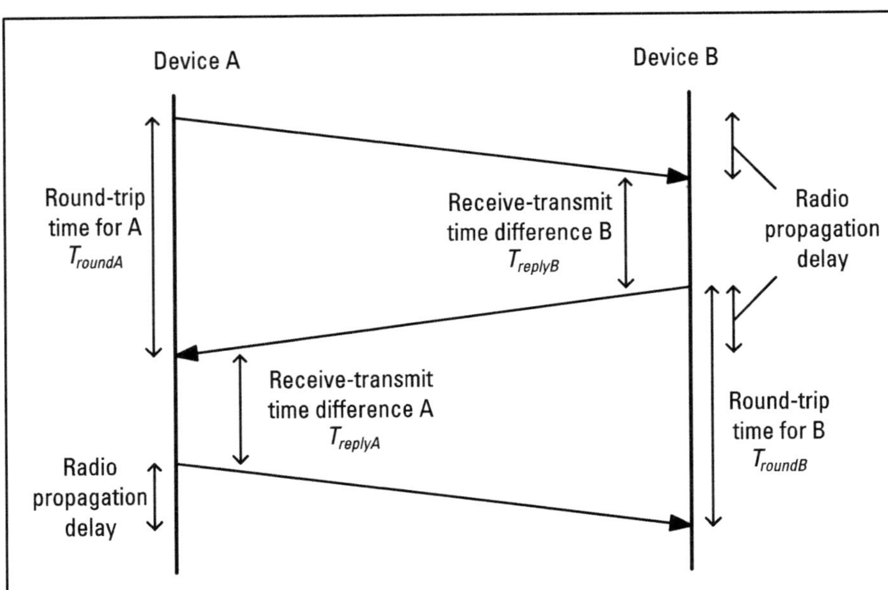

Figure 2.5 Illustration of the Ds-TWR procedure.

(3–120 cm in distance) when the reply time is 0.1 ms and clock accuracies vary from 2 to 80 ppm. However, when the reply time is increased to 5 ms, the propagation delay estimation error of the TWR significantly increases to a level of 5–200 ns (1.5–60m in distance). Note that the reply time is always limited by the used UWB packet lengths, and therefore reducing the reply time is not possible without simultaneously reducing the packet lengths. Nevertheless, as indicated before, the effect of clock frequency offset errors can be reduced by considering the Ds-TWR approach. In this case the corresponding propagation delay errors are reduced to a level of 0.005–0.2 ns (1.5–60 mm in distance) when considering the reply time of 10 ms, and the same clock accuracy range from 2 to 80 ppm.

2.1.8 High-Sensitivity GNSS

GNSS is the superior positioning system in open outdoor areas, but its use is very limited indoors. In GNSS-based positioning the traverse time of a signal from the satellite to the user receiver antenna is estimated. When this time is multiplied by the speed of light, a geometric range between the satellite and the user is obtained. In an ideal case, measurements from three satellites would provide an accurate three-dimensional position of the user. In reality the measurements are erroneous, the main error source being the timing errors between the receiver clock and the satellite clock from the system time. Therefore, the measured range is called the pseudorange. Satellite clocks are precise and synchronized by the ground control segment of the system. However, the clocks in the user receivers are low-cost with typically a large timing error. Therefore, it has to be estimated as a parameter in the navigation solution. Observations from at least four satellites are needed for three-dimensional positioning, namely the fourth observation is used for resolving the receiver clock error.

The effective isotopic radiated power (EIRP) of a GPS (L1 C/A Code) civil signal is 26.8 dBW at the time of transmission. The power decreases mainly due to free-space propagation loss (approximately 184.4 dB) while the signal travels from space to Earth. As a result, the signals are weaker than the background noise level when they reach the Earth. This fact is taken into account in the signal design, which includes a carefully selected, repeating binary code sequence modulated on the carrier; the code has a negligible autocorrelation with any delayed copy of itself. This property allows a GNSS receiver to acquire and track the satellite signals below the noise level. The receiver first generates a replica signal by predicting the code phase and carrier frequencies, and then adjusts the replica to maximize the correlation with the incoming radio samples. In order to be able to find the relevant information from the signal below noise, the minimum received power at the conventional receiver has to be around −160 dBW [56].

The requirement of the −160 dBW received power is achieved with an LOS signal outdoors, but the signal degrades due to the attenuation resulting from propagation through a material (i.e., shadowing) and interference, typically multipath (i.e., fading). When the signal penetrates building constructions, the type of the material affects the amount of attenuation. For example, while entering a concrete and steel building the mean fading of the signal ranges from 19 to 23 dB and from 12 to 21 dB, depending on the elevation angle of the satellites tracked; in worst cases, the attenuation can exceed 30 dB [57, 58]. Buildings with reflective glass walls have yet other signal attenuation properties. A technique called high-sensitivity GNSS (HSGNSS) was developed to compensate for the performance degradation indoors. The minimum received power at a typical HSGPS receiver has to be −186 dBW [57] for successful computation of the position solution. The required and received signal power levels show that the use of HSGNSS provides increased availability in most indoor environments. However, the presence of diverse materials and their arrangement indoors might result in the receiver measuring echo-only signals and thereby an erroneous solution.

In acquisition, the receiver conducts a rough two-dimensional (2D) search of code delay and Doppler frequency of the signal in order to assess whether the signal is present. In the consequent phase, tracking, the receiver follows the code delay and Doppler frequency to correctly extract the measurements needed for computing a navigation solution. Acquisition and tracking are implemented by correlation and integration. With a 2-MHZ precorrelation bandwidth and 2-ms integration time, processing gain of 33 dB is achieved. Obviously, the gain can be improved by increasing the integration time; doubling the integration period increases the sensitivity by 3 dB. However, the integration time cannot be extended arbitrarily; the receiver must be able to maintain the correct carrier frequency for the replica signal throughout the integration period. In practice this implies the following requirements:

1. The receiver needs a sufficiently stable local oscillator to mitigate clock drift;
2. The motion of the antenna needs to be known to compensate for changes in Doppler shift during the integration period.

It has been demonstrated that with a stable oven-controlled crystal oscillator and inertial sensor coupling, GNSS signals can be tracked with a coherent integration time of several seconds [59]. Nevertheless, several challenges persist for indoor GNSS, such as non-LOS signals, multipath, and the *near–far problem* where high variations in the signal strengths from different satellites cause false signal detections [58]. Although certain features of modern GNSS signals can further improve the processing gain and mitigate multipath errors, the GNSS signals will still be too heavily attenuated to be useful in many indoor environments.

2.2 Sensors

In this section we present various devices that are capable of sensing the motion or orientation of the user without relying on infrastructure such as radio signals. Some of these are completely self-contained sensors that are insensitive to external factors whereas others are dependent on environmental factors such as the ambient magnetic field or light. Many of these systems are passive (i.e., they do not emit signals that can be detected by or interfere with other sensors); however, some active imaging systems are also presented.

2.2.1 Inertial Sensors

Accelerometers and gyroscopes can be constructed by utilizing the physical inertia of a test mass, which is why they are commonly referred to as inertial sensors. They measure the linear acceleration and angular rate with respect to the I-frame. A key advantage of inertial sensing is the fact that the sensor is completely self-contained, meaning that it cannot be interfered or jammed by external effects such as magnetic fields, radio bursts, or ambient light. Furthermore, inertial sensors can be sampled at high update rates (hundreds or thousands of hertz), which makes them well suited for even high-dynamics applications.

Inertial navigation can be implemented based on a system of three accelerometers and three gyroscopes, commonly known as an inertial measurement unit (IMU). The principle is simple: acceleration can be converted to velocity, which in turn can be converted to position by integrating over time with proper initial conditions; simultaneously, the direction of travel can be tracked by integrating the gyroscope signals. In the following sections, these sensors are presented in more detail.

2.2.1.1 Accelerometers

A simple accelerometer is depicted in Figure 2.6. The test mass can move back and forth in one direction, called the sensitive axis, and is maintained at the center position by springs. Following Newton's first law, linear acceleration parallel to the sensitive axis causes the test mass to extend or contract the springs, and the deflection from the center position is proportional to the acceleration. As an alternative implementation, instead of determining acceleration from the offset of the test mass from the neutral position, the sensor can generate, for example, a magnetic force to keep the test mass at the neutral position. This principle of operation, which measures the force applied to the test mass, can give a better performance than the displacement measurement [60].

Figure 2.6(b), where the sensitive axis is oriented vertically, illustrates a fundamental property of accelerometers: they do not sense the gravitational

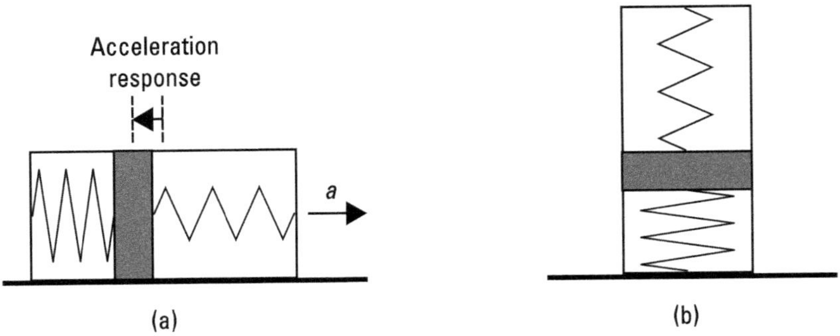

Figure 2.6 Schematic of an accelerometer consisting of a test mass mounted on springs: (a) accelerating to the right, and (b) at rest, sensitive axis upward.

acceleration. Nevertheless, they do measure the normal force exerted by the surface where the sensor is standing, which causes the output to be an upward acceleration with magnitude equal to the local gravitational acceleration (i.e., 1 g) although the sensor is at rest. This is explained by the fact that the quantity measured by an accelerometer is not exactly linear acceleration but *specific force*, which is defined as the net acceleration excluding the gravity. For instance, the specific force for an object in free fall equals zero. The specific force measurement vector f^B obtained from a triad of mutually orthogonal accelerometers can be modeled as

$$f^B = a^B - R_L^B g^L + \epsilon_f = a^B - R_L^B \begin{bmatrix} 0 \\ 0 \\ -g \end{bmatrix} + \epsilon_f \tag{2.3}$$

where the superscript B indicates a vector expressed in the *body* coordinate frame defined by the sensitive axes and L is the local level frame with axes pointing East, North, and up. The matrix R_L^B is the rotation from the L-frame to the B-frame. The vector ϵ_f denotes measurement errors that are analyzed in more detail later in this section.

In order to determine the acceleration $\boldsymbol{a}$ given the specific force f, the gravitational acceleration needs to be added to the measurement, which requires knowledge of the orientation of the sensor with respect to the gravity (i.e., the rotation R_L^B). Obviously, an accelerometer is generally unlikely to be exactly leveled (or vertical), thus the gravity compensation equals the projection of the gravitational acceleration on the sensitive axis. This calls for precise orientation information: in fact, the gravitational acceleration is a strong signal in comparison with the motion dynamics in most applications, and a 0.1° orientation error can cause a residual acceleration error of almost 0.02 m/s^2 (2 mg).

Also note that the magnitude of the gravitational acceleration g is not a global constant but depends on the position of the user. For instance, one can use the geodetic reference system 1980 gravity model, which is a function of the latitude Λ [61]:

$$g(\Lambda) = \left(1 + 0.0053024 \sin^2 \Lambda - 5.8 \times 10^{-6} \sin^2 (2\Lambda)\right) \times 9.780327 \text{ m/s}^2. \tag{2.4}$$

This model applies for the mean sea-level altitude.

On the other hand, if the accelerometer is *known* to be stationary, its output can be used to determine the orientation with respect to the gravity (i.e., the roll and pitch angles, by substituting $\boldsymbol{a}^B = \boldsymbol{0}$ in (2.3)). This way, an accelerometer can serve as an inclinometer. Note that this does not yield a unique rotation matrix R_L^B but the solution is ambiguous for rotations about the local vertical direction (i.e., the heading angle). The accuracy of the inclination estimate is driven by the measurement errors $\boldsymbol{\epsilon}_f$ following the same relationship as above: a measurement bias of 2 mg corresponds to an inclination error of 0.1°.

2.2.1.2 Gyroscopes

For tracking the orientation of the accelerometers, which is necessary for the gravity compensation, gyroscopes are commonly used. Traditionally, they were based on a spinning test mass: following the law of the conservation of angular momentum, the spin axis tends to maintain its orientation when subject to an external torque. Although mechanical gyroscopes can be very accurate, the construction is impractical for many applications: to gain a sufficient angular momentum, the test mass must be large, heavy, or spinning at a very high speed.

Most gyroscopes available in the market are either optical or microelectromechanical system (MEMS) devices. Optical gyroscopes measure the Sagnac effect between two laser beams traveling the same path in opposite directions, which can be implemented by mirrors in a ring laser gyroscope (RLG) or a coil of optical fiber in a fiber-optic gyroscope (FOG). In contrast, MEMS gyroscopes sense rotations by measuring the Coriolis acceleration of a pair of vibrating test masses; the test mass system may resemble a tuning fork or a wine glass. The principle is illustrated in Figure 2.7. The test masses are driven to vibrate in a plane parallel to the sensitive axis. When an input rotation is applied, a Coriolis force perpendicular to the driven vibration and the rotation rate vectors acts on the test masses, deflecting their motion out of the plane. The rotation rate can be resolved by measuring the deflection.

As mentioned earlier, gyroscopes measure rotation with respect to an inertial frame I; the measurement of a gyroscope triad is conveniently denoted by the angular rate vector $\boldsymbol{\omega}_{BI}^B$ expressing the rotation of the B-frame with respect to the I-frame, expressed in the B-frame. In order to model rotations with respect to the local level frame L, which is necessary for updating the rotation matrix R_B^L, we also need to account for the rotation of the L-frame itself with respect to the I-frame. Let us first factor the rotation matrix R_B^L to express rotations with

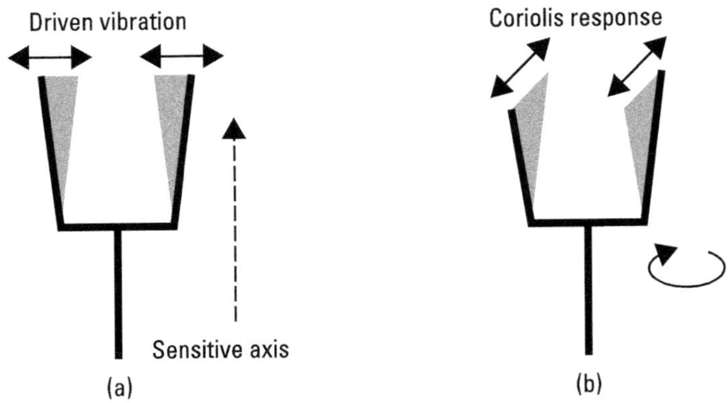

Figure 2.7 Operating principle of a tuning fork gyroscope: (a) the test masses (fork tines) are driven to vibrate on the fork plane, and (b) under rotation, the Coriolis force deflects the vibration out of the fork plane.

respect to the Earth-centered Earth-fixed frame E and the I-frame:

$$\mathrm{R}_B^L = \mathrm{R}_E^L \mathrm{R}_I^E \mathrm{R}_B^I. \tag{2.5}$$

We can differentiate this product with respect to time using the relationship [62]

$$\dot{\mathrm{R}}_A^B = \left[\boldsymbol{\omega}^B\right]_\times \mathrm{R}_A^B = \mathrm{R}_A^B \left[\boldsymbol{\omega}^A\right]_\times \tag{2.6}$$

where $[\boldsymbol{\omega}]_\times$ is the cross product matrix of the rotation rate vector, given by

$$[\boldsymbol{\omega}]_\times = \begin{bmatrix} 0 & -\omega_3 & \omega_2 \\ \omega_3 & 0 & -\omega_1 \\ -\omega_2 & \omega_1 & 0 \end{bmatrix}. \tag{2.7}$$

Now, differentiating (2.5) and applying (2.6) yields the differential equation [63]

$$\dot{\mathrm{R}}_B^L = \mathrm{R}_B^L \left[\boldsymbol{\omega}_{BI}^B\right]_\times - \left[\boldsymbol{\omega}_{EI}^L + \boldsymbol{\omega}_{LE}^L\right]_\times \mathrm{R}_B^L \tag{2.8}$$

where two rate vectors contribute to the rotation of the L-frame in addition to the gyroscope output $\boldsymbol{\omega}_{BI}^B$. First, the Earth rotates with respect to the I-frame at rate $\Omega_E \approx 15.041$ deg/h, and the resulting rotation vector can be expressed in the ENU L-frame as

$$\boldsymbol{\omega}_{EI}^L = \Omega_E \begin{bmatrix} 0 \\ \cos\Lambda \\ \sin\Lambda \end{bmatrix} \tag{2.9}$$

where Λ is the geographical latitude. Second, when the user is moving, the local vertical direction (defined by the gravitational acceleration) changes following the curvature of the Earth; this rotation is called the *transport rate* $\boldsymbol{\omega}_{LE}$ and is a function of the user's velocity.

In the context of indoor navigation, the area of operation is typically so limited that the curvature of the Earth becomes insignificant. Therefore, in the remainder of this book, we neglect the transport rate $\boldsymbol{\omega}_{LE}$. Also recall that the Earth rotation rate Ω_E is a very weak signal and may be buried under measurement errors when using low-cost gyroscopes. Moreover, compensating for the Earth rotation requires knowledge of the heading with respect to North. For these reasons, it may become practical to ignore the Earth rotation rate $\boldsymbol{\omega}_{EI}$ as well.

2.2.1.3 Characterization of Inertial Sensors

The outputs of inertial sensors are usually integrated, often several times, to obtain the navigation solution. Therefore, when selecting sensors for a navigation system, it is important to understand the different error components that affect the sensor output. The most important parameters for inertial sensors are presented in the following. The orders of magnitude are given for sensors in batch production without individual calibration; it is possible to calibrate the sensors to obtain a better performance, but this obviously increases the cost.

Bias: The additive bias corresponds to the mean value of the sensor output when no signal is present. For accelerometers, the bias limits the roll/pitch determination accuracy; for a batch-manufactured MEMS accelerometer, the bias uncertainty is typically several mg. In the context of low-cost gyroscopes, the biases are typically so large (in the order of 1°/s) that they need to be estimated and compensated for during operation. Biases can vary with temperature.

Bias instability: The bias instability quantifies the correlated short-term variations in the sensor output, giving a limit for how accurately the sensor bias can be determined. The bias instability is driven by a noise component with $1/f$ spectrum in the sensor output [64]. It is no longer uncommon for a MEMS gyroscope to reach a bias instability in the order of 1°/h.

Scale factor: Multiplicative errors are quantified by the scale factor (or sensitivity) error. For a low-cost inertial sensor, the scale factor error is typically in the order of 1%. Scale factor errors can also vary with temperature.

Random walk: Uncorrelated (white) measurement noise can be quantified in terms of angle random walk (ARW) (for gyroscopes) or velocity random walk (VRW) (for accelerometers). Random walk is expressed as the standard deviation of the integral of the noise over (the square root of) a certain period (e.g., in units of $°/\sqrt{\mathrm{h}}$ for gyroscopes or $\mathrm{m/s}/\sqrt{\mathrm{h}}$ for accelerometers). Note that the value of random walk is independent of the sensor sampling frequency.

Misalignment: The imperfect alignment of the sensitive axes during manufacturing and assembly causes two kinds of errors: misalignment of the sensors with respect to the printed circuit board (PCB) or housing they are installed to, and nonorthogonality errors resulting in cross-axis sensitivity effects. With low-cost sensors, the misalignment errors are typically expressed as a percentage of the sensor output (instead of the physical misalignment angle) and the typical order of magnitude is 1%.

As an alternative to random walk, noise can be quantified in terms of a noise density, the difference being that the noise density corresponds to the standard deviation of the error in the raw signal while the random walk quantifies the integral of the error. The noise density expresses the noise standard deviation as a function of the sampling rate (e.g., in units of $°/\text{s}/\sqrt{\text{Hz}}$) for gyroscopes because longer sampling intervals allow for filtering to reduce the measurement noise. With the assumption of uncorrelated noise, a numerical integration of a random variable sampled at f_s hertz, with standard deviation following the noise density σ_D $°/\text{s}/\sqrt{\text{Hz}}$, over one hour (h) yields the relationship:

$$\text{ARW}\left[°/\sqrt{\text{h}}\right] = \sqrt{\sum_{i=1}^{3600\text{s}/\text{h}f_s}\left(\sigma_D\sqrt{f_s}\times\frac{1}{f_s}\right)^2} = 60\sigma_D\left[°/\text{s}/\sqrt{\text{Hz}}\right], \quad (2.10)$$

that is, $0.01°/\text{s}/\sqrt{\text{Hz}}$ gyroscope noise density corresponds to $0.6°/\sqrt{\text{h}}$ ARW.

Random walk and bias instability parameters can be determined using an Allan deviation (ADEV) graph. ADEV (or equivalently its square (i.e., the Allan variance)) was originally developed for the analysis of precise clocks and is a function of averaging time τ. It is recommended to compute the ADEV using overlapping samples instead of the classical formula [65]; given a time series of N measurements y_i, $i = 1, \ldots, N$ from a stationary inertial sensor, the overlapping ADEV is computed efficiently as follows:

1. Integrate the sequence of N measurements y_i once, with zero initial condition, to obtain another time series of $N + 1$ angle or velocity values x_i.
2. Compute [65]:

$$\text{ADEV}(\tau) = \sqrt{\frac{1}{2(N-2m+1)\tau^2}\sum_{i=1}^{N-2m+1}(x_{i+2m} - 2x_{i+m} + x_i)^2} \quad (2.11)$$

   where $m = \tau f_s$ is the number of sensor samples within the averaging period τ. Obviously, the averaging period must be chosen such that m is an integer.

Conventionally, ADEV is visualized as a log–log graph with the averaging period on the x-axis. As an example, the ADEV for a certain MEMS gyroscope is plotted in Figure 2.8. The ADEV curve has a distinctive U-shape that allows us to quantify some of the error components. First, the descending slope at the left side of the plot corresponds to white noise. The standard deviation of noise decreases in proportion to the square root of the time, which is why the white noise line has a slope of $-1/2$ on the log–log graph. The noise density is obtained as the value of the white noise line at $\tau = 1$s; in Figure 2.8, we have noise density $5°/\mathrm{h}/\sqrt{\mathrm{Hz}}$ corresponding to ARW $0.09°/\sqrt{\mathrm{h}}$ (assuming the noise is perfectly white). Second, the flat region at the bottom of the curve is due to $1/f$ noise. The bias instability is defined as the minimum ADEV, which is just above $1°/\mathrm{h}$ in the example case.

It is possible to identify further error components from the ADEV curve, but they are not as commonly quoted in sensor data sheets. For instance, the rising slope at the right side of the figure is related to drifting biases, and quantization errors can be distinguished from random walk with very precise sensors. For a complete description of ADEV analysis, the reader is referred to [66].

2.2.2 Magnetometers

Probably the most common means for orientation determination is a compass where a leveled magnetic needle aligns itself with the local magnetic field. A *magnetometer* is a sensor that measures the strength of the ambient magnetic

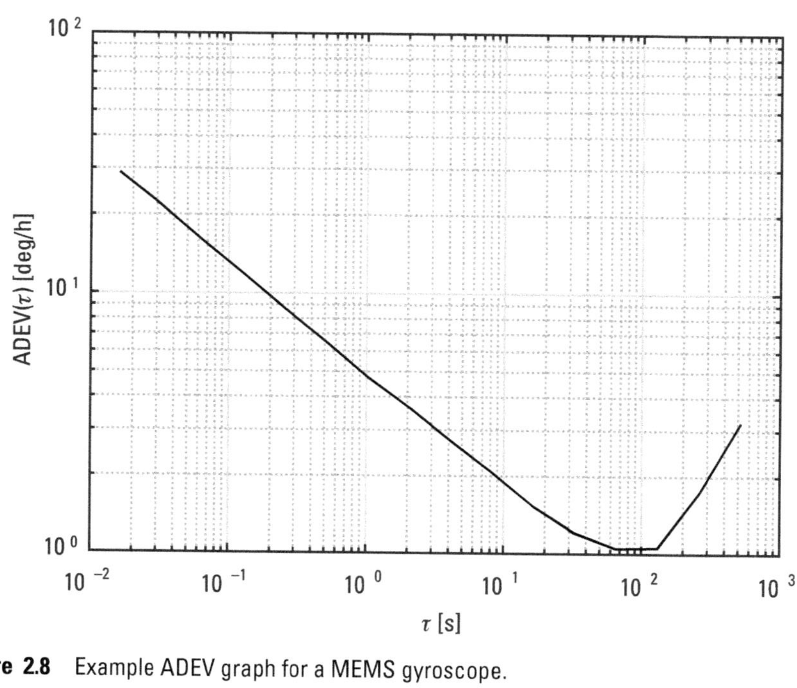

Figure 2.8 Example ADEV graph for a MEMS gyroscope.

field along a single axis, and a triad of magnetometers can be used to implement a digital compass. MEMS magnetometers are usually based on the magnetoresistive effect with the sensing element built of a metal alloy whose electrical resistivity depends on the ambient magnetic field. A better performance can be obtained using a fluxgate sensor that uses a coil and electric current to drive a test mass to magnetic saturation: an external magnetic field will affect the necessary current to saturate the mass, which becomes observable by alternating the polarity of the current.

Unfortunately, the magnetic field of the Earth is not uniform. The local magnetic field is conventionally characterized by three components:

Inclination: the angle between the field vector and the local horizontal plane (also referred to as the magnetic *dip*);
Declination: the angle between the field vector and the geographical North direction;
Intensity: the magnitude of the field vector.

Equivalently, the field can be expressed in East, North, and vertical components. In order to determine the heading with respect to the geographic North, a model of the magnetic field is needed to compensate for the magnetic declination (i.e., the offset between the magnetic and geographic North). For instance, one can use the World Magnetic Model [67], which is based on spherical harmonics and covers the entire globe, or a regional model with a more refined resolution.

The intensity of the magnetic field of the Earth varies between 23–67 microteslas at the surface of the Earth [67]. This signal can be dominated by local anomalies arising from, for example, ferromagnetic materials, power lines, and other man-made objects. Unfortunately, such anomalies are abundant in indoor environments—even the steel bars embedded in reinforced concrete can affect magnetometer measurements. This degrades the usefulness of magnetometers for indoor navigation considerably. It is noteworthy that magnetic distortions may also be caused by equipment carried by the user, such as in first responder operations.

In some applications, the susceptibility to local anomalies can be mitigated by calibration. Conventionally, magnetic field distortions are divided into two categories. *Hard iron* anomalies refer to objects that produce magnetic fields themselves, such as loudspeakers. In contrast, *soft iron* distortions only stretch or deflect the ambient field; such effects can arise from nearby ferromagnetic objects that are not permanently magnetized, such as steel. Including hard and soft iron effects, the three-axis magnetometer measurement $\boldsymbol{m}^B$ can be modeled as [68]

$$\boldsymbol{m}^B = S\mathrm{R}_L^B \boldsymbol{M}^L + \boldsymbol{h} + \boldsymbol{\epsilon}_m \tag{2.12}$$

where S denotes the soft iron distortion, $\boldsymbol{M}^L$ is the Earth magnetic field vector, and $\boldsymbol{h}$ is the hard iron anomaly; $\boldsymbol{\epsilon}_m$ represents measurement noise. A calibration algorithm should estimate the values of S and h. Various approaches have been proposed, such as [68–70].

While the unpredictable nature of the indoor magnetic field makes magnetometer-based heading determination difficult, it is possible to exploit the temporal stability of these local magnetic anomalies for position determination: Although a single magnetometer measurement would be useless, position information can be extracted from a time series of magnetometer measurements made by a moving user. For instance, one can first survey the magnetic anomalies of a building by walking around and recording the route on a floor plan, which yields a magnetic fingerprint database. Then, the magnetometer measurements of subsequent users can be matched to the fingerprint database [71]. This method can determine the user position in absolute coordinates, but the fingerprint database is tedious to create and maintain up to date. Alternatively, a SLAM approach can be taken to construct the magnetic map on the fly [72]. However, SLAM is limited to relative positioning unless other measurements are available.

2.2.3 Barometers

The ambient air pressure decreases as one moves higher in altitude. Therefore, measuring the pressure with a barometer gives access to height information, which is very important for navigation in multifloor buildings. Miniature barometers based on MEMS technology are widely available and readily integrated in various devices such as many smartphones. A MEMS barometer typically uses a membrane as the sensing element: one side of the membrane is in contact with open air while the other side faces a hermetic cavity. The force exerted by the ambient air pressure causes the membrane to deform, and the displacement can be measured, for example, capacitively.

The air pressure $p(h)$ measured at altitude h can be related to the pressure at reference height h_0 by [73]:

$$p(h) = p(h_0) \exp \frac{-gM_0 (h - h_0)}{RT_0} \tag{2.13}$$

where $M_0 = 28.9644$ kg/kmol denotes the molar mass of air, $R = 8.31432 \times 10^3 \frac{\text{Nm}}{\text{kmol K}}$ is the universal gas constant, and T_0 is the temperature at the reference height. Equation (2.13) assumes that the *temperature lapse rate* is zero (i.e., the temperature does not change with altitude). While the temperature is colder at higher altitudes in outdoor conditions, this is not necessarily the case indoors—in fact, it could actually be the opposite as warm air tends to rise up. However, in typical indoor navigation applications the height differences are so small that the effect of temperature lapse is insignificant. As a rule of thumb, one can calculate that a pressure difference of 100 pascals corresponds approximately to a height difference of 8 meters.

The key challenge involved in the use of barometric altimetry is the dependence on local atmospheric conditions (i.e., the reference pressure $p(h_0)$ and temperature T_0). These values can change relatively fast, which implies that barometers need timely reference information in addition to the calibration of the sensor itself. In indoor spaces, the ambient pressure and temperature do not follow the weather conditions because of factors such as heating and air conditioning. When implementing an indoor navigation system, these challenges can be solved by deploying reference barometers as part of the local infrastructure.

2.2.4 Optical Sensors and Systems

Here, we will discuss different optical systems used in navigation, both indoors and outdoors. Each system has its own strengths and challenges, and choosing the most suitable one among them depends on the navigation environment, performance requirements, and the boundaries set for the system, such as cost, size, and power consumption.

Optical sensors used for navigation are roughly divided into four categories according to the cameras used: single-camera methods (monocular camera), stereo cameras, red green blue-distance (RGB-D) methods, and light detecting and ranging (lidar).

2.2.4.1 Monocular Cameras

Single- (also called monocular) camera methods use a single optical camera that measures visible light or another portion of the spectrum of electromagnetic radiation. The use of a single camera for navigation is hampered by the scale problem, as will be discussed in Section 2.3. Without additional information, a single camera will not be able to solve the distance traveled on the correct scale.

Most single-camera methods use cameras that detect visible light and display images as either red, green, and blue channels, or a single grayscale channel. The quality and features of the image sensors are crucial in terms of how much noise they produce in the images. Monocular cameras have two important parameters: focal length and f-number. The focal length (f) defines the distance between the center of the lens and the film while taking a focused image of an object that is infinitely far. The focal length influences the sharpness of the image. Wide-angle lens cameras—those with a lens with a short focal length—produce sharper images than those with a standard lens. The aperture is the lens' diaphragm opening inside the camera and its diameter is presented as D in Figure 2.9. Aperture regulates the amount of light passing to the sensor and its size is indicated by an *f-number* computed as $f - number = f/D$. A large *f-number* implies a small aperture (D), leading to reduced light throughput and an increased depth of field. A smaller f-number indicates that more light is let in and there is a higher

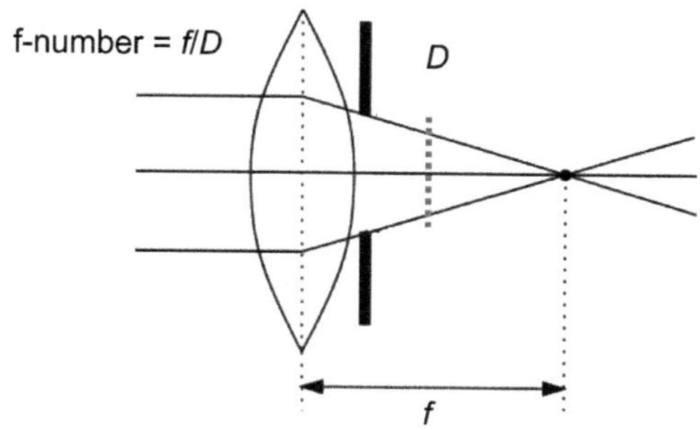

Figure 2.9 Lens, focal length, and aperture and the equation for measuring the f-number.

image quality in low-light situations. Figure 2.9 shows the lens, focal length, and aperture.

2.2.4.2 Thermal Cameras

Use of normal monocular cameras is limited to situations where there is enough light and no fog or smoke to blur the view. Thermal cameras are cameras that, instead of visible light (wavelength from 390 to 700 nm), detect infrared radiation (700 to 14,000 nm), and thus are also able to operate in the dark. The decline in the price of thermal cameras in recent years has accelerated research in their use to generate situational awareness, for example through identification of objects (people, obstacles) and imaging of the environment. It has been observed that a thermal camera works equally well in light and dark. However, feature detection and descriptor matching discussed later in this chapter show poor performance and reduced accuracy with thermal cameras and therefore more sophisticated algorithms must be used [74].

Thermal cameras have entered into indoor navigation research very recently. One representative navigation example is given in [75], where a thermal camera was integrated with other sensors into an aerial robot for navigating inside an industrial premise and in a dark subterranean mine in the presence of heavy airborne dust. The system provided real-time processing and good navigation performance when using entropy information to selectively utilize visual and thermal image information for an aerial robot's pose estimation. In general, thermal images offer lower contrast in comparison to visible light images, especially when imaging areas containing objects of similar temperature and emission values. Therefore, use of a thermal camera requires the use of algorithms improving the contrast, such as histogram equalization [76].

2.2.4.3 Stereo and RGB-D Cameras

Stereo cameras have two synchronized cameras in a rig; that is, a device for mounting two cameras rigidly together with a known distance and position. The distance to objects in the environment (depth) is calculated searching for the same object from images obtained from two cameras in the rig and using triangulation. Calculation is based on a concept called disparity, which will be explained in more detail in Section 2.3.5. However, use of a stereo camera is problematic in feature-poor environments as it is difficult or even impossible to find similar characters from the images. In addition, the distance between the two cameras, the baseline, must be long enough to obtain a working positioning solution. The error in the observed object's depth z is directly proportional to the baseline b as $\delta z \approx z^2/b$, but also depends on the image resolution. Therefore, accurate depth requires a large baseline, a large focal length, and objects being nearby, namely low depth. Roughly, this means that for a stereo rig with a 10-cm baseline and cameras with 35-mm focal length, depth observations from a 10m distance would contain 5 cm of error. Therefore, stereo cameras on, for example, mobile phones are not very feasible for navigation [77]. Thermal cameras have also been used in a stereo setup to provide visibility in low-lighting situations with metric navigation solution [78].

All the above optical sensors are passive sensors. In practice the passiveness means that the systems do not interfere with each other and also have lower power consumption as they are not transmitting anything and merely just perceiving the surroundings. An RGB-D camera is able to solve the depth information by using active range measurements like laser or infrared with only one camera. Time-of-flight cameras are active systems that measure the distance to a target by monitoring the phase difference between the light transmitted from its own light-source emitter and returning light [79]. The operating principle is very similar to the operating method of lidar. Various models of RGB-D cameras have appeared to the market, such as Microsoft Kinect, Structure IO, ASUS Xtion Pro, and Intel RealSense, and attracted a lot of attention from the navigation community due to their low cost, weight and power consumption, small size, and capability of solving the scale ambiguity. However, the use of the low-cost RGB-D cameras in navigation is still challenging due to the low quality of the depth images.

2.2.4.4 Lidar

Laser scanners measure distance using a rotating laser (or a stationary laser and a rotating mirror) [80] or optically pumped lasers in which the gain medium is solid (solid-state laser). Laser scanning is an active imaging method, as a laser beam is emitted from the scanner to perform the measurements based on the TOF method. TOF measures the round-trip flight time of the beam, where the distance to the object reflecting the beam back to the sensor is proportional to half of the ToF multiplied by the speed of light.

Laser scanning gives a three-dimensional view of the environment, a point cloud. Each point in the point cloud has x, y, and z coordinates. Based on the displacement of the point cloud features, the displacement of a moving scanner can also be calculated. Terminology related to the process is sometimes inaccurate, and the terms lidar and laser scanning are used interchangably, creating confusion. The difference between lidar technology and laser scanning is that the latter always includes determining the direction and position of the laser beam, while the former does not necessarily need it. We will continue using the term lidar as it seems to be preferred in the domain.

Most lidars have only one beam of light and they image in a two-dimensional plane. Recent years, for example with Velodyne's lidar with multiple lasers (up to 64), each scanning different levels, have allowed three-dimensional imaging. Traditionally, all lidars were 3D scanners, as the device's mirror rotated around a horizontal axis and the entire scanner remained still around a vertical axis, but in the most affordable models the rotations of the different axes are quite different. Such devices are not well suited for laser-based positioning requiring exact knowledge of the system's motion during the process. More expensive devices, on the other hand, use several lasers to measure the entire environment at a frequency of 20 Hz, which means that the distortions caused by the movement are considerably smaller and can thus be corrected by measurements from other sensors (e.g., an IMU). However, the point cloud produced in this case is less dense in the vertical direction, especially at longer distances. At present, solid-state lidars are emerging as a replacement to the traditional mechanical ones. They are built entirely on a silicon chip without any moving parts, and appear to be more resilient to vibrations, as well as being smaller and less expensive than the traditional mechanical ones.

Lidars are widely used in robotics, autonomous vehicles, and surveying because they provide accurate location information in an easy-to-understand format. In indoor navigation they have so far mainly been used for robots and mining due to the large size, weight, and processing power requirements setting restrictions to the platform. Prices for laser scanners range from a few hundred euros (low quality and unreliable equipment) to 100,000 euros (high quality professional equipment). Reliable equipment to be used in robotics and vehicles usually cost between € 1,000 and € 10,000. It is expected that the prices of laser scanners will decrease significantly in the near future, as the increasing use of scanners in vehicles will increase the production volume of the scanners used in them and thus the production costs.

Figure 2.10 presents a comparison of the characteristics of different optical systems. The comparison is done for good-quality optical sensors providing sufficient navigation performance. Systems in all classes may be obtained with lower price and size, but it would affect their usability in navigation.

2.2.5 Future Trends

During the past decade, MEMS technology has advanced at an immense pace. This development is not expected to slow down in the near future; for instance,

	Monocular cameras	Thermal cameras	RGB-D	Stereo	LiDAR
Power consumption	low	low	low	low	high
Price	reasonable	reasonable	reasonable	medium	expensive
Size	small	small	small	large	large
Tolerance for poor lighting, texture-poor areas	No	Yes	No	No	Yes
Can estimate the depth	No	No (monocular), Yes (stereo)	Yes	Yes	Yes

Figure 2.10 Characteristics of different optical systems.

mass-produced MEMS accelerometers are approaching bias tolerances of 1 m*g* over lifetime, allowing a leveling precision of 0.05°.

Simultaneously, MEMS gyroscopes are becoming stable enough to observe the Earth's rotation rate if the constant biases can be calibrated. Measuring the Earth's rotation rate would allow the possibility of absolute heading measurements; drifting heading estimates are often a bottleneck in indoor navigation systems. The Earth's rotation rate is such a weak signal that the gyroscope signals need to be averaged for some time to mitigate measurement noise. Thus, although gyroscope-based North seeking may remain unfeasible for pedestrian navigation systems, it would be possible for vehicle-based systems (e.g., in mines).

Currently, optical gyroscopes represent the state of the art in navigation systems. Sensors based on atom interferometry are at a relatively low technology readiness level (TRL) at the moment, but show great potential for high-precision sensing; some results outside laboratory conditions have been obtained [81]. Eventually they may outperform optical sensors, but it is unlikely that they would reach a cost suitable for the mass market even if they could be manufactured at chip scale; current implementations range from 10 cm to 10m in size [82].

Development trends in optical system hardware seem to be focused on solid-state lidar, which provides higher resolution than traditional lidar and is still cheaper and faster. As the sensors are less expensive to be manufactured, their prices are predicted to decrease to around 100 euros. At the time of writing this book, iPhone smartphones include a solid-state lidar with a vertical cavity surface-emitting laser (VCSEL) light source that can measure distances up to several meters with high accuracy. However, the low-cost VCSEL lidars are only suitable for smartphones that typically do not demand high range, field of view (FoV), or points per second (PPS). For autonomous systems requiring significantly higher levels of performance and reliability, higher-end and larger VCSELs must be used. For very safety-critical applications with varying distances of objects to be measured, different degrees of the light sources are required, edge-emitting lasers (EEL) are required in addition to VCSELs. Thereby, the trend is in developing methods combining VCSEL and EEL solid-state lidars for obtaining performance improvements with low cost and small size.

Related to all optical systems is the trend to develop of deep learning based measurement processing methods to correct the measurement errors arising from the sensor and setup as well as obtaining more understanding from perception. These methods will be discussed throughout this book.

2.3 Computer Vision

Computer vision incorporates various mathematical techniques allowing a computer to obtain information about the images according to human perception. Computer vision has been an active research area since the 1970s, but still in 2022 the level of detail and understanding of causality of the state-of-the-art computer vision methods remain below the level of a 2-year-old child [83]. Computer vision methods span from low-level computer vision, such as filtering out noise by using image-processing techniques, to the most recent innovations in deep learning based object understanding [84].

Computer vision has been used for localization purposes via three different setups. In the first setup, the motion and thereby the location of moving objects may be tracked using a static camera. Such a setup could be used, for example, with surveillance cameras. The second setup is based on using a database prepared for a localization environment. The earlier prepared databases might consist of images taken of the surroundings tagged with position information, locations of fiducial markers set up to the surroundings, or a map or floor plan [85]. The absolute position of the user is provided when a match is found between a query image taken by the user and a reference image in the database [86]. One of the first such applications made for a smartphone was published in 2004 by Robertson and Cipolla [87], running the calculations on a server to which the query image was sent. Hile and Borriello [88] matched features, like corners, found from the query image, into a floor plan saved in a server. The feature matching was restricted to a certain area of the floor plan using coarse position information obtained with WLAN. Database-based vision-aiding applications provide accurate positioning, but are restricted to a certain area and require extensive preparation.

Here, we are mainly concentrating on the third and most versatile setup for using computer vision in localization. We consider a setup, where the camera is attached to the user and the task is to compute the motion of the camera, and thereby the user, by tracking the change of the location of the image points representing static real-world points. In this case, a camera is seen as an additional motion sensor. Computer vision techniques in this setting entered navigation research in the 1980s. First they were developed for robots [89], then for vehicles

[90], recently for pedestrians [91] and drones [92], and currently they are at the core of autonomous navigation.

The change of the location of the image points may be observed by tracking salient and repeatable features over consecutive image frames or by looking at the *optical flow*, which is the change of the intensity values of all image pixels or of a certain area. We will start by discussing these two methods. Then, we will look at the principles of image formation and related coordinate frames. The intrinsic parameters of the camera will define how the motion of the image points relates to the detection of the camera motion, and therefore we will discuss camera calibration providing the information. The motion of the camera, more specifically the rotation and translation of the camera between two consecutive images, is related to the motion of the points at the image plane by a geometry called epipolar geometry. Epipolar geometry and its use for resolving the motion will be discussed next. A computer sees images as two-dimensional matrices of digitized brightness value functions. It is not able to detect perspectives from the images or fill the occluded parts in the scene as humans do. Therefore, it is not straightforward to expand the motion measurements between two images into complete camera motion information, and we will present some of the most important methods enabling vision-based navigation. Finally, we will discuss indoor navigation's specific challenges and enablers for vision.

2.3.1 Feature Detection and Matching

In computer vision, the 3D real-life matters that are seen in the field-of-view of the camera are called objects and their 2D images are called features. Navigation via computing the motion of the camera requires tracking of the change of the image point locations representing static objects. Features (i.e., points of interest presenting rich image content information and distinguishable across all image areas) need to be detected and matched from image to image. Feature detection and matching may be divided into three stages: (1) feature detection, where an image is searched for points that are likely to match well in other images, (2) feature description, where a region around a detected feature is converted into a more compact and stable descriptor that can be matched against other descriptors, and (3) matching, where a distance function comparing two descriptors is computed and a match declared when its value is below a set threshold. The three important concepts in the process are the image, presented throughout this book with I, the location of a pixel x,y, and the intensity value of the pixel $I(x,y)$. When detecting features, pixel values are compared and the final results of matching features are the corresponding pixel locations.

Objects in the image are separated from the background by occluding contours [93], where image brightness values change rapidly and cause large gradients. Feature detection is based on detecting these gradient values. It would be tempting to track the whole edges of the objects, for example doors or furniture in an indoor office environment. Edges of the object may be detected by algorithms such as the Canny edge detector [94], which calculates magnitudes and directions of the gradients. However, larger edge sections—for example, lines—are sensitive for occlusions. Therefore, features are usually defined as single points in the image, called interest or key points. As single points are not really distinguishable and thereby easily matched over images, descriptors also defining a certain area around the feature points are needed. Object corner points have been much used as features due to their well distinguishable nature.

The Harris corner detector [83] is one of the most famous corner detectors. A corner is detected by looking at a patch centered on an image point. The patch is then compared to nearby, overlapping patch sifted (i,j) pixels from the original patch. The comparison is done by computing the difference between the pixel values inside the two patches using a function called the sum of squared differences (SSD). A low value of SSD indicates that the two neighboring areas are similar, so not very distinctive. A corner in turn produces a large SSD value

$$SSD = \sum_{(x,y)\in W} [I(x+i, y+j) - I(x,y)]^2. \tag{2.14}$$

Figure 2.11 is an intuitive presentation of the use of the SSD function and illustrates the distinctive nature of the corners. In Figure 2.11(a) the image patch is laid over a flat region, a region in the image that represents a plane with uniform

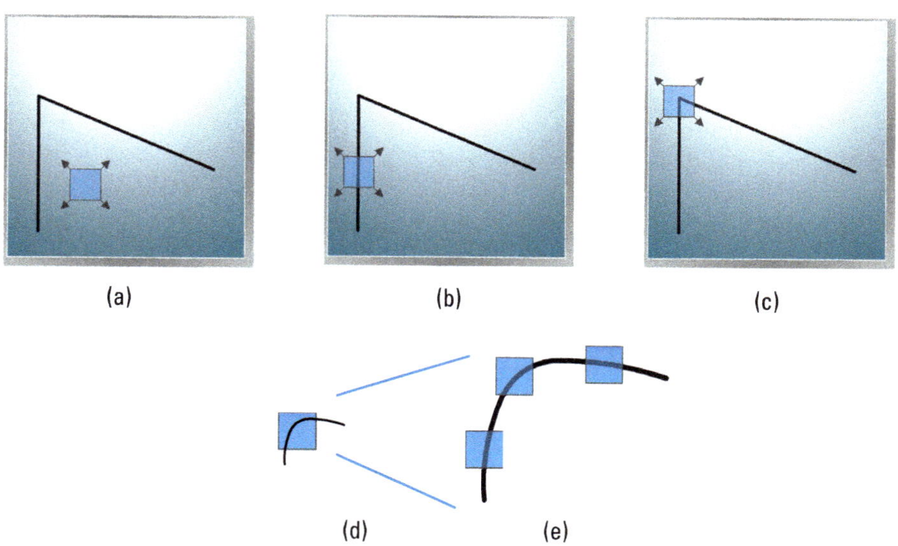

Figure 2.11 Detecting features: plane (a), edge (b), and corner (c). Scaling disturbs feature detection by changing the appearance of an object in an image such as (d) corner into (e) edges.

texture. Moving such a patch does not make large changes to the pixel values and therefore SSD is small. In Figure 2.11(b) the patch is laid over a line. In this case, moving the patch along the line does not noticeably change the pixel values and results in small SSD, while moving it perpendicular to the line provides a larger SSD. In Figure 2.11(c), moving the patch in any direction will result in a large SSD value, and therefore the corner is a good feature to be used.

The expressions of objects in images are sets of points of digitized brightness value functions, where their forms change in relation to the pose of the camera and the brightness of the environment. The appearance of a corner remains when the camera rotates or translates, and it is invariant to lighting changes to some extent. However, when the camera gets closer to the corner the size of the corner changes and therefore changes its appearance to look more like the lines seen in Figure 2.11(d).

The requirement of completely invariant features has motivated the research on scale-invariant detectors and descriptors. Scale-invariant feature transform (SIFT) [95] is an approach based on transforming an image into local feature vectors, SIFT descriptors, also describing the surroundings of detected image points. Each vector is invariant to image translation, scaling, and rotation, and partially invariant to illumination changes. SIFT provides good performance and has inspired researchers to find improved variants, such as real-time detectors FAST [96] and SURF [97]. However, these detectors use a detection process that smooths relevant detail such as object boundaries from the images while removing noise. When navigating in a feature-poor environment, as many indoor environments are, it is essential to detect all features, even the weaker ones. Kaze [98] is a feature detector and descriptor based on nonlinear diffusion filtering and therefore performs better than, for example, SIFT, in challenging indoor scenes. Kaze detects features that have a high distinctiveness, but is unfortunately slow in computation, 2.5 times slower than SURF, for example.

Matching is a process of identifying the corresponding features in two images taken at different viewpoints, different times, or by different cameras. It is implemented by comparing pairwise each feature descriptor computed for the two images and using a distance function defining the similarity. The most probable match is found at the minimum of the distance function. Euclidean distance (L_2- norm) is an often-used metric, defined for a two-dimensional vector as $d(p,q) = \sqrt{(q_1 - p_1)^2 + (q_2 - p_2)^2)}$, and a matching threshold is defined to discard too-weak candidates. In many cases brute-force matching is used, which simply takes the descriptor of one feature in the first image and matches it with all other descriptors in the second image.

After successfully matching features between two images, there is a set of n image point coordinates $\mathbf{x}_{i1} = (x_{i1}, y_{i1}), \mathbf{x}_{i2} = (x_{i2}, y_{i2}), i = 1, \ldots, n$ from which the camera motion between the epochs of taking the two images may be estimated as explained in Section 2.3.3.

Feature-based motion tracking performs well in texture-rich environments, such as indoor furnished rooms. However, various textureless or low-textured indoor environments exist, such as long corridors and stairways seen in Figure 2.12, where feature-based methods fail completely. The tolerance for feature-poor environments is still an open research question; however, the optical flow methods discussed next might provide some answers to the problem.

2.3.2 Optical Flow

Optical flow methods provide improved motion detection performance in lower repetitive-textured scenes with dynamic objects [99]. Optical flow tracks the pixel location changes over time (i.e., the pixel velocity (u, v)), in consecutive images by looking at the intensity changes of the whole image. The method is based on the assumption that the intensity of pixels representing certain real-world objects remain constant over time. Figure 2.13(a) shows the motion of a pixel in consecutive images $I(x, y, t)$ through time (t) epochs $1, \ldots, k$. The pixel represents the same real-world object point, but its image location (x,y) changes throughout the images as the camera (or object) moves. Figure 2.13(b) shows the change of the pixel location in images $I(x, y, t)$ and $I(x, y, t + \delta t)$ acquired at times t and $t + \delta t$. The assumption is called the brightness constancy constraint and it could be presented, following the principle shown in Figure 2.13, as

$$I(x + u\delta t, y + v\delta t + t + \delta t) = I(x, y, t). \tag{2.15}$$

Optical flow is computed by minimizing (2.15), which requires linearization of the equation using the multivariate Taylor series expansion [100]. Linearization

Figure 2.12 Low-textured image from office stairways.

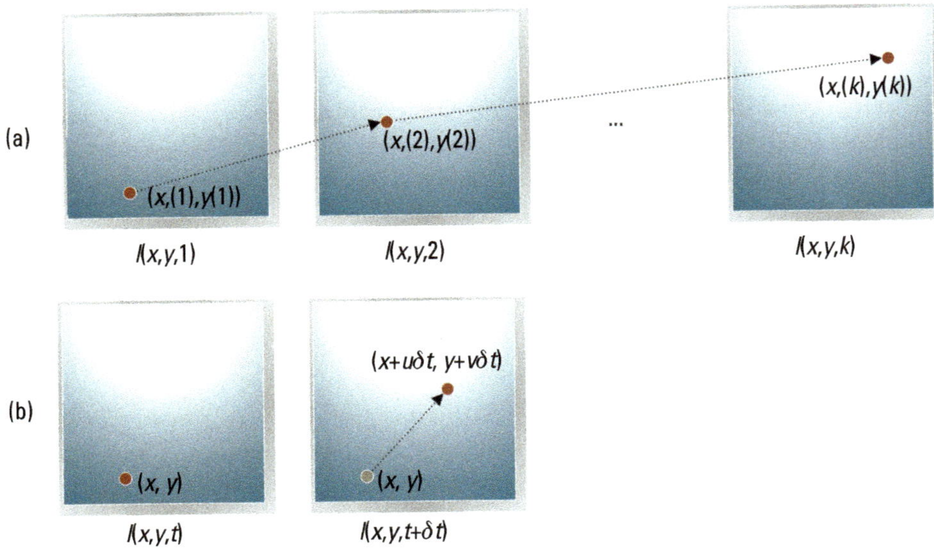

Figure 2.13 Optical flow: (a) motion of a pixel in consecutive images, and (b) change of the pixel location in images over time.

is enabled by enforcing the second assumption in optical flow: small motion. Small motion means that pixels (objects) move only a small amount between images and therefore optical flow performs best when the image rate is high. After linearization and rearranging the equation, the brightness constancy equation is

$$\frac{\partial I}{\partial x}\frac{dx}{dt} + \frac{\partial I}{\partial y}\frac{dy}{dt} + \frac{\partial I}{\partial t}\frac{dx}{dt} = 0, \tag{2.16}$$

where ∂ denotes the partial derivative and d the derivative. In a shorthand notation the brightness constancy equation is

$$I_x u + I_y v + I_t = 0, \tag{2.17}$$

where u, v are the pixel velocity components in the x and y direction, and I_x, I_y partial derivatives with respect to x and y, respectively (i.e., the image gradients), and I_t is the temporal gradient. The partial derivatives I_x, I_y will be computed via filtering, as discussed in Section 2.3.4.1, and the temporal gradient I_t by just subtracting the brightness values of corresponding pixel locations across the two images. This leaves two unknowns, u, v, the pixel velocity components, which are exactly the optical flow values. Unfortunately, for each u, v pair there is only one known pixel value. Therefore, optical flow suffers from the aperture problem; the solution might lie in any location of a line defined by the variable pair. For overcoming the lack of information and enabling computation of an unambiguous solution, different methods have been developed. The most used method is the Horn-Schunk optical flow [101], which uses a constraint called smooth flow. Smooth flow means using an assumption that most objects in the world are rigid or deform elastically and therefore their image pixels move together coherently. The Horn-Schunk optical flow algorithm uses energy minimization for optimizing the sum of the brightness constancy between temporally corresponding pixels and spatial smoothness term-regularizing neighboring pixels to have similar motion to overcome the aperture problem.

Dense optical flow based motion estimation has challenges arising from the fact that the pixel intensities also change due to illumination changes and reflection effects of certain object surfaces. Motion discontinuities, occlusions, large camera displacements, and computational costs also complicate the motion estimation and system setup [102]. Optical flow may also be computed using a sparse method. Sparse methods are a combination of feature detection and tracking of the pixel intensity changes, where pixel velocities are computed only for extracted features and not for the whole image. Sparse methods solve some of the abovementioned challenges in measuring optical flow, but at the cost of decreased accuracy.

Despite active research and development, traditional computer vision based methods still struggle with obtaining sufficient performance. Therefore, optical flow based methods have also been underrepresented in indoor navigation. However, currently research has been concentrating on the deep learning based methods discussed in Section 2.3.7, which show promising results.

2.3.3 Perspective Projection and Epipolar Geometry

The objects in the 3D world are mapped into 2D image features using perspective projections. These projections do not preserve the properties of shape, length, angle, distance, or ratio of distances, but they do preserve the property of straightness. As a result, lines that are parallel in the real world seem to intersect in an image at a virtual point, called the vanishing point. Therefore, to obtain a projective geometry space, the Euclidean geometry has to be augmented with a point and line in the infinity. The change of the geometry space causes a change in the representation of coordinates; two coordinates (x, y) presenting a point in the Euclidean space are replaced in the projective space with a triplet (x, y, 1) called homogeneous coordinates. In computer vision, the 2D image points are usually presented with homogeneous coordinates as (x,y,1) and 3D object points as (X,Y,Z,1).

2.3.3.1 Computer Vision Specific Coordinate Frames

Computer vision defines three coordinate frames, shown in Figure 2.14, two of which are additional to the ones presented in Chapter 1. The coordinate frames

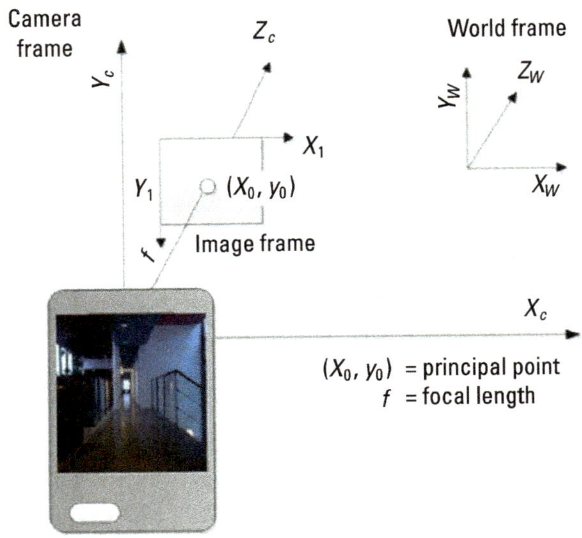

Figure 2.14 Computer vision coordinate frames: world (w), camera (c), and image (I) frames.

are called world, camera, and image frames. When using computer vision for navigation, the world frame is often aligned to the navigation frame, and its coordinates presented with homogeneous coordinates. An important note is that as opposed to the established convention of coordinate frame configuration in navigation, computer vision sets the *y*-axis pointing up and *z*-axis along the camera's principal axis. The camera frame has its origin at the optical center of the camera and axes with similar setting as the world frame, shown in Figure 2.14. The 3D world points presented in the camera frame are mapped to image frame as 2D points via the perspective projection. The image frame is a two-dimensional plane and has its origin at the left upper corner with the *y*-axis pointing down.

2.3.3.2 Camera Model and Matrix

The perspective projection is usually modeled with a simple pinhole camera model. Perspective projection maps the intersection of a ray from a 3D real-world object $\mathbf{X} = (X, Y, Z)^T$ to the camera's optical center with the image plane. The intersection point is the 2D image point in homogenous coordinates $\mathbf{x} = (x, y, 1)^T$. Projection is presented as

$$(X, Y, Z)^T \rightarrow \mathbf{x} = (fX/Z, fY/Z, f)^T, \tag{2.18}$$

where f is the focal length presented in Section 2.2.4. This projection causes one of the most significant challenges in computer vision. While turning the 3D information into 2D, perspective projection loses the information of the object's Z coordinate (i.e., the distance to the camera). The reason is that all object points lying on the ray defined by the optical center and image plane map to the same image point. This is called the depth problem and makes it impossible to solve the metric scale of the translation without any other information on constraints.

In practice, the perspective projection includes two phases and uses homogeneous coordinates for representing the object **X** and image point **x**. First, it maps the object point **X** represented in the world frame into the camera frame. The mapping is parametrized by the displacement of the camera center, presented with translation vector **t** (3×1), and its orientation with (3×3) matrix **R**, with respect to the world frame origin and axis. Then, the camera-centered 3D point X_c is mapped to 2D pixel coordinates **x**. Mapping is parametrized with a camera calibration matrix **K** discussed in more detail in Section 2.3.3.4. In its entirety, the projection mapping the object points to an image is encompassed in a camera model presented with a matrix **P**, as $\mathbf{x} = \mathbf{PX}$. Following the above discussion, the 3×4 matrix **P** may be decomposed as

$$\mathbf{P} = \mathbf{K}[\mathbf{R}|\mathbf{t}]. \tag{2.19}$$

With this we have defined the representation of an object point on an image plane. As discussed previously, to use the camera as a navigation sensor, we need to compute its motion between consecutive images. Therefore, we need to define the principle relating the camera motion and the change of the pixel locations between the consecutive images. This principle is called epipolar geometry and will be discussed next.

2.3.3.3 Epipolar Geometry

Epipolar geometry determines the relative location of image points (**x**1), ((**x**2)) representing the same real-world object (X) in two images $(I1, I2)$. Epipolar geometry defines a triangular plane, epipolar plane, as shown in Figure 2.15.

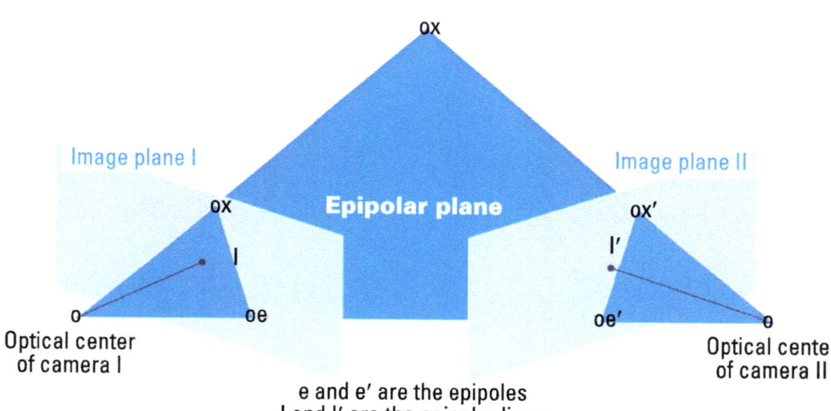

Figure 2.15 Epipolar geometry.

The corners of the plane are defined by the coordinates of the object point (X) and the optical centers of two cameras used for capturing the images as shown in Figure 2.15. The coordinates of the two camera centers are related by a displacement vector (dt) presented with translation (**t**) and rotation (**R**). The points where the image planes $I1$ and $I2$ intersect with dt are called epipoles (e1) and (e2), respectively. Image point (x1) and epipole (e1) are end points of an epipolar line (l1). Similarly, epipolar line (l2) is defined as a line spanning from (x2) to (e2).

In navigation, usually only the location of the image pixels is known and used to solve translation **t** and rotation **R**. In other applications, where translation and rotation are known, the search for features in image $I2$ matching features found from image $I1$ is constrained by the epipolar line.

In the following section, we will discuss the camera calibration matrix **K** and its formation, and then continue discussing how epipolar geometry may be used for resolving camera motion.

2.3.3.4 Camera Calibration

Navigation requires Euclidean reconstruction (i.e., image-based information that is in correct distances and angles). This may be realized by using a calibrated camera. Calibration provides information about a camera's intrinsic parameters, represented with a calibration matrix **K**. The intrinsic parameters are focal length $\mathbf{f} = (f_x, f_y)$, principal point (x_0, y_0), skew coefficient (S), aspect ratio, and distortions. Focal length f was discussed in Section 2.2.4. The principal point is the intersection of the camera's optical axis with the image plane, as shown in Figure 2.14. Distortions blur the image due to the variation of the focal length at different points of the lens. Distortions get larger when the camera's lens gets wider. The skew is a result of manufacturing errors and makes the two image axes nonorthogonal. The skew coefficient defines the angle between axes. The general form of the 3×3 matrix **K** is

$$\mathrm{K} = \begin{bmatrix} f_x & s & x_0 \\ 0 & f_y & y_0 \\ 0 & 0 & 1 \end{bmatrix}. \tag{2.20}$$

Usually, skew is neglected, and a simplified form of **K** is used:

$$\mathrm{K} = \begin{bmatrix} f_x & 0 & x_0 \\ 0 & f_y & y_0 \\ 0 & 0 & 1 \end{bmatrix}. \tag{2.21}$$

In general, it is adequate for a navigation system to calibrate the camera once and assume the parameters are unchanged thereafter. In theory, the calibration process is done by placing a known object in the scene, identifying correspondences between the image and scene and computing a mapping from scene to image. However, the process requires very accurate knowledge of the scene geometry and of the 3D to 2D correspondences. Therefore, usually calibration is done by using multiplane calibration. There, a model image, usually a chessboard pattern with measured square sizes, is photographed from different viewpoints and the parameters calculated using the known model geometry. A simple tool for performing calibration with MATLAB® is [103]. Also, methods for more automatic calibration exist, for example based on observing vanishing points from the image, from which the focal length and center of projection (the principal point) may be determined [104]. When the accuracy of the navigation solution may be compromised for the sake of adaptability, calibration may be skipped and parameters obtained straight from the image information. Focal length may be found from the image's Exchangeable Image File (EXIF) data but is an average of values the cameras of the type in question have and therefore, not as accurate as the one obtained by calibration. As well, the principal point may be assumed to be the central point of the image.

In navigation, as in most of the computer vision applications, extended field of view provides better performance. Field of view is extended by using a wide-angle lens. The downside is the radial distortion appearing in the images. In radial distortion, coordinates in the observed images are displaced away (barrel distortion) or toward (pincushion distortion) the image center by an amount proportional to their radial distance. Most computer vision algorithms assume that the projection from 3D space to the image is linear, which is true only if the image is not distorted. Therefore, radial lens distortion has to be corrected to get an optimal solution [105].

The radial distance (r_d) of the normalized distorted image points (x_d, y_d) from the radial distortion center (principal point (x_0, y_0)), is [106]

$$r_d = \sqrt{x_d^2 + y_d^2} \tag{2.22}$$

Using the radial distance of the distorted image points, the radial distance (r) of the corrected image points (x_c, y_c) is obtained as

$$r = r_d(1 - k_1 r_d^2 - k_2 r_d^4) \tag{2.23}$$

The constants k_i are the distortion values specific to the camera and are obtained via calibration. The corrected and distorted image points are related as

$$x_d = x_c(1 + k_1 r + k_2 r^2)$$
$$y_d = y_c(1 + k_1 r + k_2 r^2)$$

However, the rectification of the whole image introduces aliasing effects complicating, for example, feature detection. For optimal results, the radial distortion should be corrected only for the extracted features and not for the whole image.

2.3.3.5 Fundamental and Essential Matrices and Camera Motion

Algebraic representation of the epipolar geometry is called the fundamental matrix (**F**) [107]. For any pair of corresponding image points $(\mathbf{x1}, \mathbf{x2})$ it holds that $\mathbf{x2}^T\mathbf{F}\mathbf{x1} = 0$. This is the result of epipolar geometry; when the points **x1** and **x2** correspond, **x2** lies on epipolar line $\mathbf{l2} = F\mathbf{x1}$. From the definition it follows that the fundamental matrix encompasses the projective geometry between two views: the rotation and translation of the two camera centers and the internal camera parameters represented by the calibration matrix **K**. In practice, in navigation the camera is the same and the two center locations are related to two consecutive time epochs. Thereby, solving the fundamental matrix provides means for resolving the camera's motion between taking two images.

Fundamental matrix may be solved if at least seven image point correspondences (i.e., matching image features) are found. The most feasible algorithm for computing the fundamental matrix is called the normalized 8-point algorithm [108], and as can be guessed by its name, it requires eight point correspondences for an accurate solution.

Unfortunately, the fundamental matrix fails to preserve the true metric structure, for example the orthogonal 3D lines or planes. Therefore, a normalized version of the algebraic representation, 3×3 essential matrix (**E**), is used. Essential and fundamental matrices are related to each other via $\mathbf{E} = \mathbf{K}_2^T\mathbf{F}\mathbf{K}_1$, where $\mathbf{K}_1$ is the calibration matrix of the camera at the time of taking the first image and $\mathbf{K}_2$ the second. As we want to calculate the motion of the camera, we use the following setup. We set the camera matrix (**P1** of the first epoch to be at the world origin and aligned with the world coordinate axis $\mathbf{P1} = \mathbf{K}_1[I|0]$, where I is an identity matrix and $\mathbf{P2} = \mathbf{K}_2[\mathbf{R}|\mathbf{t}]$. Thereby, the mapping from image point **x1** to image point **x2** is

$$\mathbf{x2} = \mathbf{K}_2\mathbf{R}\mathbf{K}_1^{-1}\mathbf{x1} + \mathbf{K}_2\mathbf{t}/Z. \tag{2.24}$$

Z is the distance between the object whose image point is **x1**, so the distance between the camera and the object. Essential matrix may be computed using the same 8-point algorithm as for the fundamental matrix, but also using other algorithms requiring less correspondences, down to five [109]. The benefit of using algorithms that need fewer points for estimation is that they are less sensitive to outliers. For computing the essential matrix, image points $i = 1, 2$ must be normalized using their camera calibration matrices $\mathbf{K}_i$ as $\hat{\mathbf{x}}i = \mathbf{K}_i^{-1}\mathbf{x}i$.

After computing the essential matrix, translation and rotation are computed by decomposing **E** via singular value decomposition (SVD). SVD provides us $\mathbf{E} = \mathbf{U}\Sigma\mathbf{V}'$, from which we get **t** as the singular value (diagonal entries of σ)

associated with the smallest singular vector (columns of **U** and **V**). **R** in turn can be calculated from

$$\mathbf{R} = \pm \mathbf{U}\mathbf{R}^{T}_{\pm 90^\circ}\mathbf{V}^{T} \tag{2.25}$$

providing four possible rotation matrices. The reason for getting ambiguous results lies in the nature of SVD not being able to determine **U** and **V** uniquely from **E** for the following reasons: (1) the singular values and their corresponding singular vectors can be permuted, (2) each singular vector is unique only up to a sign, and (3) repeated singular values do not have unique singular vectors but can be rotated around.

The correct value for rotation is obtained as follows. From the four solutions the ones with $det|\mathbf{R}| = 1$, where *det* denotes the determinant, are kept. Then, the remaining solutions are paired with both possible signs of translation direction. Finally, the combination of possible rotations and translations is selected for which the largest number of object points reconstructed using the image points and **R**, **t** as shown in Figure 2.16, is seen in front of both cameras [108].

One of the biggest challenges in computer vision is that the absolute translation between the two camera positions can never be recovered from pure image measurements alone, regardless of how many image points or cameras are used. This means that the translation can be resolved only up to an ambiguous scale. Unlike in many other applications, in navigation it is unacceptable to solve

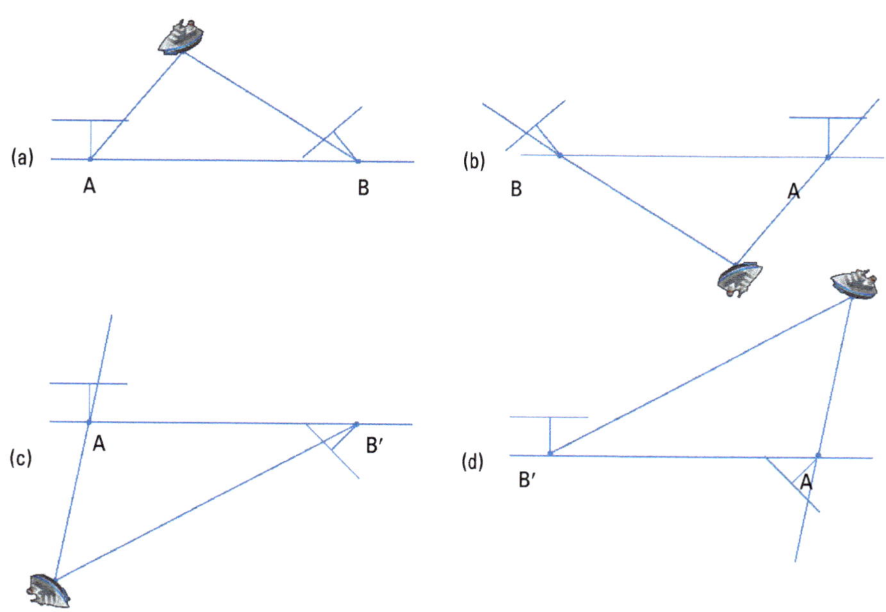

Figure 2.16 In (a), both cameras (A and B) can see the object, in (b) neither can, in (c) B sees it but A does not, and (d) B can see it but A cannot.

the motion only up to a scale as a metric solution is required. Solutions for the problem are discussed in Section 4.2.2.2.

2.3.4 Error Sources in Computer Vision

Computer vision algorithms are very error-prone. As commented throughout this chapter, perception based on computer vision is highly sensitive to *environmental conditions*. Such conditions are changes in lighting and illumination, which changes how things appear in the images, presence of shadows that form features of imaginary objects, shiny surfaces, and feature-poor areas. Navigation is also sensitive to *operational conditions*, such as occlusion, which is something appearing between the camera and the tracked object as well as dynamic objects in the view of camera confusing motion detection and resulting in an erroneous solution. The latter adds the magnitude of motion of the dynamic objects to the values calculated for the camera motion. Camera characteristics and the processing algorithms add their own errors to the solution. Here, we will discuss the main error sources in computer vision based navigation and some solutions for correcting and compensating them.

2.3.4.1 Image Formation Related Errors

The brightness of a pixel is a function of the brightness of the object's surface [93]. This analogous signal, represented with a continuous function and arriving to the camera's image sensor, is first integrated and then digitized. During this process, if the fill factor on the sensor's chip is small and the signal is sampled at less than twice the highest frequency present in the signal, aliasing will occur. Aliasing mixes high and low spatial frequencies causing erroneous brightness values for pixels in the image. Image acquisition also includes other camera- and environment-related error sources, which introduce mainly Gaussian noise into the image. The level of illumination, the camera sensor's own temperature, and the electronic circuits connected to the sensor produce electronic circuit noise [110]. Two main types of noise in the images are called salt-and-pepper and shot noise. The former will create dark pixels in bright regions and bright pixels in dark regions, while the latter forms random variations of brightness.

Especially in navigation, the motion of the camera may cause blurriness in the images. Motion blur results when the image being recorded changes during a single exposure due to camera's rapid movement or long exposure, and is an issue especially when the environment includes sources of fast flashing lights. The root cause for blur is aliasing, as discussed above. Blur appears especially when using a method called rolling shutter, which is prevalent in most consumer-grade cameras with complementary metal–oxide–semiconductor (CMOS) sensors. Another shutter type is a global shutter, which captures the entire image frame at the same instant, unlike a rolling shutter. In CMOS cameras global shutters are still quite rare, but are often found from charge-coupled device (CCD) sensors, which are generally more sensitive and more expensive.

The quality of images may be improved by using filtering. Filtering enables, in addition to removing noise in images, local tone adjustment, sharpening of image details, and accentuation of edges. It also gives tools for antialiasing, which means removing signal components that have a higher frequency than is able to be properly resolved by the sensor. Filtering is implemented using convolution. Convolution is an element-wise operation with input data and a kernel, also called a filter. Kernel (**H**) is a small ($k * k$) array of numbers (usually k is 3 or 5), that is slid across the input image (I) and at each location (i,j) calculating an element-wise product, and summing the products to produce a result image (G). Mathematically the process can be written as

$$G[i,j] = \sum_{u=-k}^{k} \sum_{v=-k}^{k} \mathbf{H}[u,v]I[i-u,j-v]. \tag{2.26}$$

Figure 2.17 shows a numerical result (image G) of applying the convolution operator with a simple mean filtering kernel **H** to an image I. Usually, the kernels are much more sophisticated and case-specific, examples being Gaussian, bandpass like Sobel, and when the noise is not Gaussian distributed, nonlinear filters, like median, are used [93].

Convolution is also the method for detecting image gradients that are used for feature extraction and getting information from images.

2.3.4.2 Algorithmic Errors

In addition to the errors in image representation, most of the algorithms used in computer vision are based on estimation and thus bring inaccuracies to the perception. Camera calibration may cause inaccurate parameters that are used throughout the computation processes and will lead to a recurrence of errors. Distortion in particular results in erroneous computation and therefore should be corrected.

Feature detection and description are delicate processes with different complications and considerations that were discussed in Section 2.3.1. One phenomenon present in indoor environments is the existence of non-Lambertian surfaces. Surfaces are called Lambertian if they reflect light uniformly toward all directions. Non-Lambertian surfaces, such as glass and metal, have complex reflectance exhibiting view-dependent characteristics that disturb feature detection. Recovering the reflectance of non-Lambertian surfaces remains a challenging problem since the view-dependent appearance invalidates traditional

1/9	1/9	1/9
1/9	1/9	1/9
1/9	1/9	1/9

H

0	0	0	0	0	0	0	0
0	0	90	90	90	90	90	0
0	0	90	90	90	90	90	0
0	0	90	90	90	90	90	0
0	0	90	0	90	90	90	0
0	0	90	90	90	90	90	0
0	0	0	0	0	0	0	0
0	90	0	0	0	0	0	0

I

=

	20	40	60	60	60	40	
	30	60	90	90	90	60	
	30	50	80	80	90	60	
	30	50	80	80	90	60	
	20	30	50	50	60	40	
	20	30	30	30	30	20	

G

Figure 2.17 Image (I) is convolved with kernel (H) to obtain a filtered image (G).

photo-consistency constraints [111] and crates complications for navigation where the view changes continuously.

Challenges in feature description are repeated in the sensitive matching process, also discussed in Section 2.3.1. Feature matching is usually accompanied with a robust estimation method called random sample consensus (RANSAC) [112]. The RANSAC algorithm extends the minimum set needed for representing a model with observed points within some error tolerance. This is done by searching for a random sample of points that leads to a fit of the desired model for which many of the points agree. This leaves the outliers out from data, for example the matched image point set. RANSAC performs well in various estimation processes and therefore the algorithm is used widely in computer vision applications.

2.3.5 Visual Odometry

Visual odometry means estimating the camera's pose (i.e., the position and orientation) over time by using only a stream of images. One or multiple cameras may be used for capturing the images; the process is also called ego-motion estimation [113]. Pose is estimated by propagating the known initial state (position and orientation) of the user with the traveled distance (translation) and direction (orientation) computed from the image flow [114]. The state may be propagated using measurements computed for each consecutive image pair independently, but a more robust solution is obtained if the information obtained over time is included. The temporal domain may be included using *bundle adjustment* or *Kalman filtering*. As errors are inevitable, the propagated distance in particular and therefore the position starts to drift, which means that absolute information is needed to reinitialize the trajectory from time to time. *Simultaneous localization and mapping* provides the means for the reinitialization. These three methods are discussed in Chapter 3. This section discusses solutions for solving the remaining issue—scale ambiguity in translation—caused by the depth uncertainty.

Various methods have been developed to solve the scale problem when using a monocular camera. So far, the most generalizable solution has been the initialization of the distance between the first two camera poses to one [115]. In this setup, when a new image arrives, the relative scale and camera pose with respect to the first two frames are estimated using a concept called trifocal tensor. The trifocal tensor is an algebraic representation of the projective geometric relationships between three views, as the fundamental matrix was for two [107]. As the geometry depends only on the relative motion of the camera, the initially set translation value may be corrected using this relationship.

In a special case, when there are objects of known size in the environment, the scale problem can be solved by using multiple image points detected from these objects [116] and the perspective projection equations discussed in Section 2.3.3.2. However, this is rarely the case in an unfamiliar environment. Similarly, there are solutions using a special setup of the navigation equipment. If the height of the camera is known and pointed straight down, the scale problem can be solved as the depth of all the objects seen in images is the height of the camera [117]. One of the most promising solutions for solving the depth issue and additionally correcting many other perception errors, is fusing the vision-based measurements with those from other positioning sensors [118]. Recently, a concept called visual-inertial odometry (VIO) has provided well-performing indoor navigation solutions [119].

As discussed previously, the use of special cameras or optical systems provides information about the depth of objects and thereby solves the scale ambiguity. RGB-D cameras and lidars provide the information using active mode, which is transmitting light or other signals that will reflect back from the objects and provide a distance measurement. Stereo cameras also provide the distance information, but their use is not as straightforward [120]. When two cameras are completely aligned, meaning their optical axes are parallel, the depth of an object may be calculated from the motion of the pixels between the images. As discussed previously, stereo means that two cameras are set in a rig with a known distance between each other, called the baseline, and known mutual orientation. Knowledge of the camera setup is obtained by calibration, enabling virtual alignment of the cameras. This process is called rectification. In rectification, first the essential matrix is computed and with it the mutual orientations of the cameras. Then, the cameras are aligned horizontally and their pitch corrected if they have been tilted downward or upward. As both cameras look ahead after alignment, the amount of horizontal motion of the features presenting the same object is inversely proportional to the object's distance from the camera. The amount of the pixel's horizontal motion is called disparity (d) and it is defined as

$$d = x - x' = \frac{baseline * f}{z}, \tag{2.27}$$

where baseline is the distance between the optical centers of the two cameras, f is the focal length, x is the image point's x-coordinate in the image taken with the first camera and x' with the second, and z is the distance to the object. Figure 2.18 shows the setup, with two cameras having centers C and C'. Thereby, z is directly proportional to the baseline and focal length. A stereo camera with a longer baseline covers objects at a longer distance, similarly when increasing the focal length. Also, choosing a longer baseline but a wide-angle lens with a shorter focal length cancels out the gain of extended baseline.

2.3.6 Indoor Navigation-Specific Features

As discussed throughout this section, the main indoor specific computer vision error sources are the varying lighting conditions of the environment, the low amount of distinctive features to be detected, and dynamic objects, mainly human. Fortunately, indoor environments also set some constraints for the setup, which might help in navigation. In this section we will discuss two of them. First, we will present vanishing points, which provide additional information for computing the motion (i.e., the orientation of the camera). Then, we will discuss navigation using landmarks. The process would not be feasible in large open outdoor environments, but is indoors due to the restricted size.

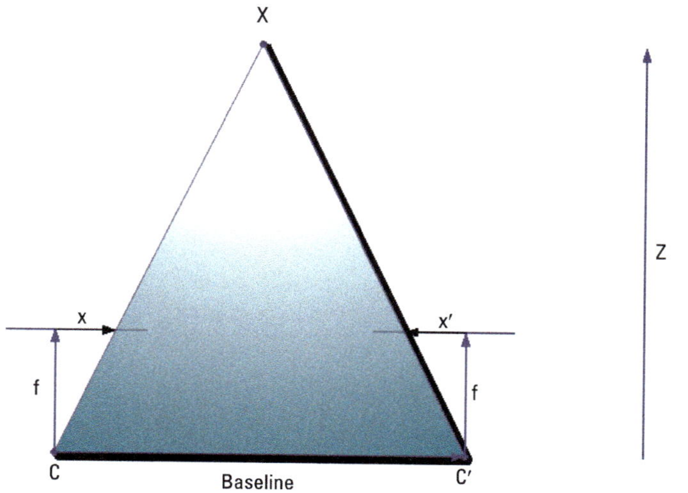

Figure 2.18 Stereo camera setup for computing the depth z by using disparity. x is the image point of the first camera C and x' of the second C'. f is the focal length.

2.3.6.1 Manhattan Assumption and Vanishing Points

Most human-made environments, such as buildings and other indoor areas, consist of segments forming a Cartesian coordinate system with straight lines in three orthogonal directions. Such a coordinate system is called the Manhattan grid [121]. A concept called the Manhattan world assumption relies on the environment to be built using the grid.

A vanishing point is the intersection point **v** of a ray through the camera center having a direction **d**, and of all other lines also having the same direction, and the image plane. The vanishing point **v** is related to the direction **d** as $\mathbf{v} = \mathbf{Kd}$ [38] where **K** is the camera calibration matrix. The directions $\mathbf{d}_1$ and $\mathbf{d}_2$ of two vanishing points in consecutive images are related by the rotation matrix **R** as $\mathbf{d}_2 = \mathbf{Rd}_1$ (i.e., rotation of the camera between the two images). The change of the position of the camera between the images, pure translation with no rotation, has no effect on the vanishing point location.

If the optical axis is set to coincide with the constructions of the environment, the central vanishing point $\mathbf{v}_z$ lies at the principal point and the other two vanishing points at infinity on the x and y image axis. Then, the orientation of the camera is described with $\mathbf{V} = \mathbf{KR}$, where **V** is the vanishing point matrix incorporating the horizontal, vertical, and central vanishing points, $[v_x v_y v_z]$ and $v_i = (x_{v_i}, y_{v_i}, 1)^T$. Now the changes from this initial orientation will change the positions of the vanishing point and provides information about our orientation change.

Methods presented in [91] compute the user attitude by monitoring the motion of the vanishing points. Lines required for computing the vanishing points are more robust for low lighting and poor-feature environments, such as an office corridor, than other features. Lines may be detected by first performing feature detection for finding edges, for example with a Canny line detector [94] and searching for lines among all edges with a Hough line detector [122]. Vanishing points may be located at the intersection points of all lines going to the same direction. The change of vanishing point locations over images resolves the user's heading, roll, and pitch. By resolving the orientation (R) the requirement for the amount of the matched feature descriptors also decreases, facilitating the estimation process.

2.3.6.2 Navigation Using Landmarks

Previous sections discussed the setup where the camera has been monitoring the motion of the user by tracking the location change of the features in images. Such methods are independent of hardware set to the environment beforehand. However, such methods suffer from various errors and, as mentioned before, provide only relative measurements with scarce possibilities to correct the solution. Many applications would tolerate the requirement of equipping the area with hardware or setting up the environment, such as navigation in public buildings. Examples of complex buildings with navigation needs are airports, railway stations, hospitals, and museums.

Artificial landmarks, for example QR codes, are called fiducials. Their use in navigation has been an active research area for decades [123]. Fiducials, when detected correctly, provide information about the user's location and orientation. Ceiling lamps have been used as fiducials [124], providing a low-cost vision localization system without the requirement to set up the environment with additional hardware. As both methods have their drawbacks, methods combining artificial and natural landmarks, such as walls, have also been investigated [125].

Another method using existing information and preparation is the use of databases of images or other material describing the environment. In the preparation phase, a building may be photographed extensively and each image tagged with a position. After creating a database of these reference images, their landmarks may be compared to the ones found from the photos captured by a navigating user, and the absolute position obtained [86, 87]. Floor plans [88] and maps may also be used as position references.

All these methods offer accurate positioning, but their restrictions are that they require an existing or a priori constituted infrastructure and are available only in a predefined environment.

2.3.7 Future Trends

Deep learning has been achieving significant results in computer vision research since the late 1990s. Recent developments in machine learning algorithms, along with availability of more processing power and valuable data, have also made deep learning based methods the main trend in computer vision based navigation research. Deep learning is a family of machine learning methods that are based on artificial neural networks [126]. Traditional deep networks are built with layers that are called fully connected. They usually process the data in one-dimensional arrays, which would lose the prevailing spatial correlation of features in images. Convolutional neural networks (CNNs) contain convolutional layers that enable processing of spatial patterns, making them the most favorable building blocks of neural networks in computer vision. In neural networks, the information between the layers is transferred using a function (y)

$$y = \Sigma_{i=1}^{n} w_i x_i + b, \tag{2.28}$$

where x_i is the value of input i, w_i is its weight, n is the total number of inputs and b is a bias (i.e., threshold value). In supervised deep learning, the actual learning task is to find the w and b values so that the model is able to join the input with the preferred output. In the training phase the network is given the correct input and corresponding output data and after learning the model should be able to do the predictions for unprecedented data.

Convolution operation is exactly the same as in the case of using more traditional image processing such as filtering the images. In Figure 2.17, the values of the kernel (**H**) were entered manually; in CNNs the values are the weights learned in the process. Here, convolution is used for detecting features from images. The first layers of a CNN learn primitive features, such as lines, and throughout the layers these features add up to more representative objects.

Discussed in Section 3.7, the state-of-the-art indoor navigation solutions are based on neural networks. As noted, low-level computer vision tasks, like feature detection and basic algorithms, such as optical flow, are increasingly being based on CNNs.

2.4 Summary

In this chapter, we studied various types of measurements used in positioning and navigation systems. First, we discussed systems requiring specific setup of infrastructure, focusing fully on radio systems, including cellular network technologies from 2G to 5G and beyond, Wi-Fi, Bluetooth, UWB, and high-sensitivity GNSS. We then studied sensors carried by the user without requiring external measurement infrastructure, and detailed considering inertial sensors, magnetometers, barometers, and optical systems.

Regarding radio systems, we presented different positioning-related measurements available in the considered radio standards from mobile networks to Bluetooth, Wi-Fi, and UWB systems, while also highlighting the characteristics of high-sensitivity GNSS. We discussed related system-specific requirements, challenges, and limitations regarding achievable measurement accuracy, while also considering the essential characteristics of each radio standard from a positioning perspective.

From radio systems we proceeded to study various types of sensors, potentially available in the user device. First, we focused on inertial sensors, including accelerometers and gyroscopes, which compose the main components of inertial navigation through IMU. We listed the most significant error sources affecting the sensor outputs. After this, we proceeded to sensors whose outputs are dependent on environmental factors, considering magnetometers for observing ambient magnetic field, barometers for observing ambient air pressure, and optical sensors for detecting light. From these, magnetometers contribute to indoor positioning by capturing specific building-originated local perturbation in Earth's magnetic field, while barometers provide valuable information for elevation and floor estimation. Finally, we studied optical sensors, including single camera methods, stereo cameras, RGB-Ds, and lidars, while discussing the related implementation challenges. We provided rough guidelines for system selection, and additionally clarified the terminology used with optical systems.

The remaining part of the chapter was dedicated to computer vision for mapping optical measurements of a chosen camera setup into user motion. Related to this, we studied characteristics, advantages, challenges, and significant error sources of different optical systems, and further highlighted some specific features of related indoor navigation systems. We studied methods of feature matching and optical flow for motion estimation, and additionally, visual odometry for pose estimation. In addition, we considered aspects of camera calibration, discussed mapping of the 3D world into 2D image features using perspective projection, and presented the concept of epipolar geometry.

In the future, the continuous development of radio standards reveals new opportunities for improved positioning performance, specifically in terms of accuracy, latency, and security. Beyond 5G, mobile networks are expected to offer improved range and angle estimation accuracy through increased bandwidth and more directive antenna arrays along with other emerging technologies such as reflective intelligent surfaces. Furthermore, machine learning methods are expected to become increasingly popular in solving complicated positioning and SLAM scenarios. Besides radio systems, machine learning methods, especially CNNs, are increasingly being utilized in computer vision systems, for example, with feature matching and optical flow methods.

References

[1] European Telecommunications Standards Institute (ETSI), TS 145005, *Digital Cellular Telecommunications System (Phase 2+) (GSM); GSM/EDGE Radio Transmission and Reception*, May 2022.

[2] European Telecommunications Standards Institute (ETSI), TS 145001, *Digital Cellular Telecommunications System (Phase 2+) (GSM); GSM/EDGE Physical Layer on the Radio Path; General Description*, May 2022.

[3] European Telecommunications Standards Institute (ETSI), TS 144 031, *Digital Cellular Telecommunications System (Phase 2+) (GSM); Location Services (LCS); Mobile Station (MS) – Serving Mobile Location Centre (SMLC) Radio Resource LCS Protocol (RRLP)*, May 2022.

[4] European Telecommunications Standards Institute (ETSI), TR 138901, *5G; Study on Channel Model for Frequencies from 0.5 to 100 GHz*, 2017.

[5] Sun, S., T. S. Rappaport, T. A. Thomas, et al., "Investigation of Prediction Accuracy, Sensitivity, and Parameter Stability of Large-Scale Propagation Path Loss Models for 5G Wireless Communications," *IEEE Transactions on Vehicular Technology*, Vol. 65, No. 5, 2016, pp. 2843–2860.

[6] Lee, B.-H., D. Ham, J. Choi, S.-C. Kim, and Y.-H. Kim, "Genetic Algorithm for Path Loss Model Selection in Signal Strength-Based Indoor Localization, *IEEE Sensors Journal*, Vol. 21, No. 21, 2021, pp. 24285–24296.

[7] Rodriguez, I., H. C. Nguyen, I. Z. Kovacs, T. B. Sorensen, and P. Mogensen, "An Empirical Outdoor-To-Indoor Path Loss Model from Below 6 gHz to cm-wave Frequency Bands," *IEEE Antennas and Wireless Propagation Letters*, Vol. 16, 2017, pp. 1329–1332.

[8] Kumar, C., and K. Rajawat, "Dictionary-Based Statistical Finger printing for Indoor Localization," *IEEE Transactions on Vehicular Technology*, Vol. 68, No. 9, 2019, pp. 8827–8841.

[9] Ibrahim, M., and M. Youssef, "Cellsense: An Accurate Energy-Efficient GSM Positioning System," *IEEE Transactions on Vehicular Technology*, Vol. 61, No. 1, 2012, pp. 286–296.

[10] Ergen, S. C., H. S. Tetikol, M. Kontik, R. Sevlian, R. Rajagopal, and P. Varaiya, "RSSI-Fingerprinting-Based Mobile Phone Localization with Route Constraints," *IEEE Transactions on Vehicular Technology*, Vol. 63, No. 1, 2014, pp. 423–428.

[11] European Telecommunications Standards Institute (ETSI), *TS 125 215, Universal Mobile Telecommunications System (UMTS); Physical Layer; Measurements (FDD)*, 2020.

[12] European Telecommunications Standards Institute (ETSI), *TS 125 225, Universal Mobile Telecommunications System (UMTS); Physical Layer; Measurements (TDD)*, 2020.

[13] European Telecommunications Standards Institute (ETSI). *TS 125 133, Universal Mobile Telecommunications System (UMTS); Requirements for Support of Radio Resource Management (FDD)*, 2018.

[14] European Telecommunications Standards Institute (ETSI), *TS 125 123, Universal Mobile Telecommunications System (UMTS); Requirements for Support of Radio Resource Management (TDD)*, 2018.

[15] European Telecommunications Standards Institute (ETSI), *TS 125 305, Universal Mobile Telecommunications System (UMTS); Stage 2 Functional Specification of User Equipment (UE) positioning in UTRAN*, May 2022.

[16] Dahlman, E., S. Parkvall, and J. Skold, *4G, LTE-Advanced Pro and the Road To 5G*, San Diego, CA: Elsevier Science & Technology, 2016.

[17] European Telecommunications Standards Institute (ETSI), *TS 136 211, LTE; Evolved Universal Terrestrial Radio Access (E-UTRA); PhysicalChannels and Modulation*, August 2021.

[18] European Telecommunications Standards Institute (ETSI), *TS 136 104, LTE; Evolved Universal Terrestrial Radio Access (E-UTRA); Base Station (BS) Radio Transmission and Reception*, 2018.

[19] European Telecommunications Standards Institute (ETSI), *TS 136 101, LTE; Evolved Universal Terrestrial Radio Access (E-UTRA); User Equipment (UE) Radio Transmission and Reception*, 2020.

[20] European Telecommunications Standards Institute (ETSI), *TS 136 305, LTE; Evolved Universal Terrestrial Radio Access Network (E-UTRAN); Stage 2 Functional Specification of User Equipment (UE) Positioning in E-UTRAN*, 2021.

[21] European Telecommunications Standards Institute (ETSI), *TS 121 915, Digital Cellular Telecommunications System (Phase 2+) (GSM); Universal Mobile Telecommunications System (UMTS); LTE; 5G;* Release Description; Release 15, October 2019.

[22] Holma, H., A. Toskala, and T. Nakamura, *5G Technology*, John Wiley & Sons, 2020.

[23] European Telecommunications Standards Institute (ETSI), *TS 138 211, 5G; NR; Physical Channels and Modulation*, 2020.

[24] European Telecommunications Standards Institute (ETSI), *TS 138 101-1, 5G; NR; User Equipment (UE) Radio Transmission and Reception; Part 1: Range 1 Standalone*, November 2020.

[25] European Telecommunications Standards Institute (ETSI), *TS 138 305, 5G; NG Radio Access Network (NG-RAN); Stage 2 Functional Specification of User Equipment (UE) Positioning in NG-RAN*, April 2021.

[26] European Telecommunications Standards Institute (ETSI), *TS 138 133, 5G; NR; Requirements for Support of Radio Resource Management*, 2021.

[27] Bourdoux, A., A. Noll Barreto, B. van Liempd, et al., *6g White Paper on Localization and Sensing*, 2020.

[28] Mazuelas, S., A. Bahillo, R. M. Lorenzo, et al., "Robust Indoor Positioning Provided by Real-Time RSSI Values in Unmodified WLAN Networks," *IEEE Journal of Selected Topics in SignalProcessing*, Vol. 3, No. 5, 2009, pp. 821–831.

[29] Talvitie, J., M. Renfors, and E. S. Lohan, "Distance-Based Interpolation and Extrapolation Methods for RSS-Based Localization with Indoor Wireless Signals, *IEEE Transactions on Vehicular Technology*, Vol. 64, No. 4, 2015, pp. 1340–1353.

[30] Feng, C., W. S. A. Au, S. Valaee, and Z. Tan, "Received-Signal-Strength-Based Indoor Positioning Using Compressive Sensing," *IEEE Transactions on Mobile Computing*, Vol. 11, No. 12, 2012, pp. 1983–1993.

[31] Cao, H., Y. Wang, J. Bi, S. Xu, H. Qi, M. Si, and G. Yao, "WiFi RTT Indoor Positioning Method Based on Gaussian Process Regression for Harsh Environments," *IEEE Access*, Vol. 8, 2020, pp. 215777–215786.

[32] Guo, G., R. Chen, F. Ye, X. Peng, Z. Liu, and Y. Pan, "Indoor smartphone Localization: A Hybrid WiFi RTT-RSS Ranging Approach," *IEEE Access*, Vol. 7, 2019, pp. 176767–176781.

[33] Makki, A., A. Siddig, and C. J. Bleakley, "Robust High Resolution Time of Arrival Estimation for Indoor WLAN Ranging," *IEEETransactions on Instrumentation and Measurement*, Vol. 66, No. 10, 2017, pp. 703–2710.

[34] Sun, M., Y. Wang, L. Huang, H. Jia, J. Bi, W. Joseph, and D. Plets, "Geomagnetic Positioning-Aided Wi-Fi FTM Localization Algorithm for NLOS Environments," *IEEE Communications Letters*, Vol. 26, No. 5, 2022, pp. 1022–1026.

[35] LAN/MAN Standards Committee, IEEE Computer Society, *IEEE Std802.11-2020, Part 11: Wireless LAN Medium Access Control (MAC) and Physical Layer (PHY) Specifications*, December 2020.

[36] Chen, C., C. Yan, H. Yi, L. Hung-Quoc, Z. Feng, and K. R. Liu, "Achieving centimeter-accuracy indoor localization on WiFi Platforms: A Multi-Antenna Approach," *IEEE Internet of Things Journal*, Vol. 4, No. 1, 2017, pp. 122–134.

[37] He, S., and S.-H. G. Chan, "Wi-Fi Fingerprint-Based Indoor Positioning: Recent Advances and Comparisons," *IEEE Communications Surveys Tutorials*, Vol. 18, No. 1, 2016, pp. 466–490.

[38] Chen, Z., H. Zou, J. F. Yang, H. Jiang, and L. Xie, "WiFi Fingerprinting Indoor Localization Using Local Feature-Based Deep LSTM," *IEEE Systems Journal*, Vol. 14, No. 2, 2020, pp. 3001–3010.

[39] Guo, X., S. Zhu, L. Li, F. Hu, and N. Ansari, "Accurate WiFi Localization by Unsupervised Fusion of Extended Candidate Location Set," *IEEE Internet of Things Journal*, Vol. 6, No. 2, 2019, pp. 2476–2485.

[40] Wang, X., L. Gao, S. Mao, and S. Pandey, "CSI-Based Fingerprinting for Indoor Localization: A Deep Learning Approach," *IEEE Transactions on Vehicular Technology*, Vol. 66, No. 1, 2017, pp. 763–776.

[41] Banin, L., O. Bar-Shalom, N. Dvorecki, and Y. Amizur, "Scalable Wi-Fi Client Self-Positioning Using Cooperative FTM-Sensors," *IEEE Transactions on Instrumentation and Measurement*, Vol. 68, No. 10, 2019, pp. 3686–3698.

[42] Ma, C., B. Wu, S. Poslad, and D. R. Selviah, "Wi-Fi RTT Ranging Performance Characterization and Positioning System Design," *IEEE Transactions on Mobile Computing*, Vol. 21, No. 2, 2022, pp. 740–756.

[43] Shao, W., H. Luo, F. Zhao, H. Tian, S. Yan, and A. Crivello, "Accurate Indoor Positioning Using Temporal—Spatial Constraints Based on Wi-Fi Fine Time Measurements," *IEEE Internet of Things Journal*, Vol. 7, No. 11, 2020, pp. 11006–11019.

[44] IEEE Computer Society, *IEEE P802.11az, IEEE Draft Standard for Information Technology - Telecommunications and information Exchange Between Systems Local and Metropolitan Area Networks - Specific Requirements Part 11: Wireless LAN Medium Access Control (MAC) and Physical Layer (PHY) Specifications - Amendment 4: Enhancements for Positioning.*

[45] Bluetooth SIG, Core Specification Working Group, Bluetooth Core Specification, 2021.

[46] Munir, M. S., D. H. Kim, A. K. Bairagi, and C. S. Hong, "When CVAR Meets with Bluetooth PAN: A Physical Distancing System for Covid-19 Proactive Safety," *IEEE Sensors Journal*, Vol. 21, No. 12, 2021, pp. 13858–13869.

[47] Su, Z., K. Pahlavan, and E. Agu, "Performance Evaluation of Covid-19 Proximity Detection Using Bluetooth LE Signal," *IEEE Access*, Vol. 9, 2021, pp. 38891–38906.

[48] Gentner, C., D. Gunther, and P. H. Kindt, "Identifying the BLE Advertising Channel for Reliable Distance Estimation on Smartphones," *IEEE Access*, Vol. 10, 2022, pp. 9563–9575.

[49] Nikodem, M., and P. L. Szelinski, "Channel Diversity for Indoor Localization Using Bluetooth Low Energy and Extended Advertisements," *IEEE Access*, Vol. 9, 2021, pp. 169261–169269.

[50] Bluetooth SIG, Enhancing Bluetooth Technology; Key Specification Projects, 2022.

[51] ITU-R. Rec. ITU-R SM.1755-0 1, Characteristics of Ultra-Wideband Technology, 2006.

[52] Coppens, D., E. De Poorter, A. Shahid, S. Lemey, and C. Marshall, "An Overview of Ultra-Wideband (UWB) Standards (IEEE802.15.4, FIRa, Apple): Interoperability Aspects and Future Research Directions," 2022.

[53] LAN/MAN Standards Committee, IEEE Computer Society, *IEEE802.15.4-2020 Standard for Low-Rate Wireless Networks*, May 2020.

[54] LAN/MAN Standards Committee, IEEE Computer Society, *IEEE Standard 802.15.4z-2020 for Low-Rate Wireless Networks, Amendment1: Enhanced Ultra Wideband (UWB) Physical Layers (PHYs) and Associated Ranging Techniques*, 2020.

[55] IEEE P802.15 Working Group for Wireless Personal Area Networks (WPANs), *Application of IEEE Std 802.15.4*, May 2014.

[56] Kaplan, E., and D. Hegarty, *Understanding GPS: Principles and Applications*, Norwood, MA: Artech House, 2006.

[57] Lachapelle, G., H. Kuusniemi, D. Dao, G. Macgougan, and M. Cannon, "HSGPS Signal Analysis and Performance Under Various Indoor Conditions," *Navigation*, Vol. 51, No. 1, 2004, pp. 29–43.

[58] Seco-Granados, G., J. A. Lopez-Salcedo, D. Jimenez-Banos, and G. Lopez-Risueno, "Challenges in Indoor Global Navigation Satellite Systems: Unveiling its Core Features in Signal Processing," *IEEE Signal Processing Magazine*, Vol. 29, No. 2, 2012, pp. 108–131.

[59] Pany, T., J. Winkel, B. Riedl, et al., "Coherent Integration Time: The Longer, the Better," *Inside GNSS*, November–December 2009, pp. 52–61.

[60] Eren, H., "Acceleration, Vibration, and Shock Measurement," in *Measurement, Instrumentation, and Sensors Handbook* (J. G.Webster, ed.), CRC Press, 1998, pp. 553–586.

[61] Helmut, M., "Geodetic Reference System 1980," *Journal of Geodesey*, Vol. 62, No. 3, 1988, pp. 348–358.

[62] Zhao, S., "Time Derivative of Rotation Matrices: A Tutorial," arXiv:1609.06088v1, September 2016.

[63] Savage, P. G., "Computational Elements for Strapdown Systems," RTO Educational Notes RTO-SET-116(2008), NATO Research and TechnologyOrganization, 2009.

[64] Keshner, M. S., "1/f Noise," *Proceedings of the IEEE*, Vol. 70, March 1982, pp. 212–218.

[65] Riley, W. J., *Handbook of Frequency Stability Analysis*, NIST Special Publication 1065, Boulder, CO: National Institute of Standards and Technology, July 2008.

[66] IEEE Standard Specification Format Guide and Test Procedure for Coriolis Vibratory Gyros, IEEE Std 1431-2004, Institute of Electrical and Electronics Engineers, 2004.

[67] Chulliat, A., W. Brown, P. Alken, et al., *The US/UK World Magnetic Model for 2020–2025: Technical Report*, National Centers for Environmental Information, NOAA, 2020, doi:10.25923/ytk1-yx35.

[68] Vasconcelos, J. F., G. Elkaim, C. Silvestre, P. Oliveira, and B. Cardeira, "Geometric Approach to Strapdown Magnetometer Calibration in Sensor Frame," *IEEE Transactions on Aerospace and Electronic Systems*, Vol. 47, No. 2, 2011, pp. 1293–1306.

[69] Renaudin, V., M. Haris Afzal, and G. Lachapelle, "Complete Triaxis Magnetometer Calibration in the Magnetic Domain," *Journal of Sensors*, December 2010.

[70] Wu, Y., and W. Shi, "On Calibration of Three-Axis Magnetometer," *IEEE Sensors Journal*, Vol. 15 No. 11, 2015, pp. 6424–6431.

[71] Haverinen, J., and A. Kemppainen, "Global Indoor Self-Localization Based on the Ambient Magnetic Field," *Robotics and Autonomous Systems*, Vol. 57, No. 10, 2009, pp. 1028–1035.

[72] Jung, J., T. Oh, and H. Myung, "Magnetic Field Constraints and Sequence-Based Matching for Indoor Pose Graph SLAM," *Robotics and Autonomous Systems*, Vol. 70, 2015, pp. 92–105.

[73] U.S. Standard Atmosphere, 1976, Technical Report NOAA–S/T 76–1562, National Oceanic and Atmospheric Administration, Washington, D.C., October 1976.

[74] Mouats, T., N. Aouf, D. Nam, and S. Vidas, "Performance Evaluation of Feature Detectors and Descriptors Beyond the Visible," *Journal of Intelligent and Robotic Systems*, Vol. 92, 2018, pp. 1–31.

[75] Khattak, S., C. Papachristos, and K. Alexis, "Visual-Thermal Landmarks and Inertial Fusion for Navigation in Degraded Visual Environments," in *2019 IEEE Aerospace Conference*, IEEE, 2019, pp. 1–9.

[76] Naik, S. K., and C. A. Murthy, "Hue-Preserving Color Image Enhancement without Gamut Problem," *IEEE Transactions on Image Processing*, Vol. 12, No. 12, 2003, pp. 1591–1598.

[77] Holzmann, C., and M. Hochgatterer, "Measuring Distance with Mobile Phones Using Single-Camera Stereo Vision," in *2012 32nd International Conference on Distributed Computing Systems Workshops*, Macau, China, 2012, pp. 88–93.

[78] Mouats, T., N. Aou, L. Chermak, and M. A. Richardson, "Thermal Stereo Odometry for UAVS," *IEEE Sensors Journal*, Vol. 15, No. 11, 2015, pp. 6335–6347.

[79] Huang, A., A. Bachrach, P. Henry, M. Krainin, D. Maturana, D. Fox, and N. Roy, "Visual Odometry and Mapping for Autonomous Flight Using an RGB-D Camera," in *Robotics Research: The 15th International Symposium ISRR*, Springer International Publishing, 2017, pp. 235–252.

[80] Pekkala, J., *3d-laserkeilausaineiston hyödyntäminen inframallintamisen yhteydessä ja sen lopputuotteen laadun varmentaminen*, Master's Thesis, Tampere University of Technology, 2005.

[81] Bongs, K., M. Holynski, J. Vovrosh, et al., "Taking Atom Interferometric Quantum Sensors from the Laboratory to Real-World Applications," *Nature Reviews Physics*, Vol. 1, 2019, pp. 731–739.

[82] Geiger, R., A. Landragin, S. Merlet, and F. Pereira Dos Santos, "High-Accuracy Inertial Measurements with Cold-Atom Sensors," *AVS Quantum Science*, Vol. 2, 2020.

[83] Szeliski, R., *Computer Vision: Algorithms and Applications*, Second Edition, Springer, 2022.

[84] Zhang, C., Z. Cui, Y. Zhang, B. Zeng, M. Pollefeys, and S. Liu, "Holistic 3D Scene Understanding from a SingleImage with Implicit Representation," in *Proceedings of the IEEE/CVF Conference on Computer Vision and Pattern Recognition*, 2021, pp. 8833–8842.

[85] Aoki, H., B. Schiele, and A. Pentland, "Realtime Personal Positioning System for Wearable Computers," in *Proceedings of the 3rd IEEE ISWC*, 1999.

[86] Zhang, W., and J. Koseck, "Image Based Localization in Urban Environments," in *The Third International Symposium on 3D Data Processing, Visualization and Transmission*, IEEE Computer Society, 2006, pp. 33–40.

[87] Robertson, D., and R. Cipolla, "An Image-Based System for Urban Navigation, in *Proceedings of the British Machine Vision Conference*, London, UK, 2004, pp. 260–272.

[88] Hile, H., and G. Borriello, "Information Overlay for Camera Phones in Indoor Environments," in *3rd International Symposium on Location and Context Awareness, Lecture Notes in Computer Science*, Vol. 4718, 2007, pp. 68–84.

[89] Bonin-Font, F., A. Ortiz, and G. Oliver, "Visual Navigation for Mobile Robots: A Survey," *Journal of Intelligent and Robotic Systems*, Vol. 53, No. 3, 2008, pp. 263–296.

[90] Campbell, N. W., M. R. Pout, M. D. J. Priestly, E. L. Dagless, and B. T. Thomas, "Autonomous Road Vehicle Navigation," *Engineering Applications of Artificial Intelligence*, Vol. 7, No. 2, 1994, pp. 177–190.

[91] Ruotsalainen, L., M. Kirkko-Jaakkola, J. Rantanen, and M. Makela, "Error Modelling for Multi-Sensor Measurements in Infrastructure-Free Indoor Navigation," *Sensors*, Vol. 18, No. 2, 2018.

[92] Al-Kaff, A., D. Martín, F. García, A. de la Escalera, and J. M. Armingol, "Survey of Computer Vision Algorithms and Applications for Unmanned Aerial Vehicles," *Expert Systems with Applications*, Vol. 92, 2017, pp. 447–463.

[93] Forsyth, D., and J. Ponce, *Computer Vision: A Modern Approach*, Second Edition, Pearson, 2012.

[94] Canny, J. F., "A Computational Approach to Edge Detection," *IEEE Transactions on Pattern Analysis and Machine Intelligence*, Vol. 8, No. 6, 1986, pp. 79–698.

[95] Lowe, D. G., "Distinctive Image Features from Scale-Invariant Keypoints," *International Journal of Computer Vision*, Vol. 60, No. 2, 2004, pp. 91–110.

[96] Rosten, E., and T. Drummond, "Machine Learning for High-Speed Corner Detection," in *Computer Vision–ECCV 2006: 9th European Conference on Computer Vision*, Graz, Austria, May 7–13, 2006, Springer, pp. 430–443.

[97] Bay, H., T. Tuytelaars, and L. Gool, "SURF: Speeded Up Robust Features," in *Computer Vision–ECCV 2006: 9th European Conference on Computer Vision*, Graz, Austria, May 7–13, 2006, Springer, pp. 404–417.

[98] Alcantarilla, P. F., A. Bartoli, and A. J. Davison, "Kaze Features," in *Computer Vision—ECCV (Lecture Notes in Computer Science)*, Vol. 7577, 2012, pp. 214–227.

[99] Xu, J., R. Ranftl, and V. Koltun, "Accurate Optical Flow via Direct Cost Volume Processing," in *Proceedings of the IEEE Conference on Computer Vision and Pattern Recognition*, Honolulu, HI, 2017, pp. 1289–1297.

[100] Thomas, G. B., and R. L. Finney, *Calculus and Analytic Geometry*, 9th Edition, Reading, MA: Addison Wesley, 1996.

[101] Horn, B. K. P., and B. G. Schunck, "Determining Optical Flow," *Artificial Intelligence*, Vol. 17, No. 1, 1981, pp. 85–203.

[102] Fortun, D., P. Bouthemy, and C. Kervrann, "Optical Flow Modeling and Computation: A Survey," *Computer Vision and Image Understanding*, Vol. 134, 2015, pp. 1–21.

[103] Bouguet, J.-Y., Camera Calibration Toolbox for MATLAB, http://www.vision.caltech.edu/bouguetj/calib_doc/.

[104] Kosecka, J., and W. Zhang, "Video Compass," in *Proceedings of the European Conference on Computer Vision*, 2002, pp. 657–673.

[105] Ruotsalainen, L., J. Bancroft, G. Lachapelle, H. Kuusniemi, and R. Chen, "Effect of Camera Characteristics on the Accuracy of a Visual Gyroscope for Indoor Pedestrian Navigation," in *2012 Ubiquitous Positioning, Indoor Navigation, and Location Based Service (UPINLBS)*, 2012, pp. 1–8.

[106] Ma, L., Y. Chen, and K. Moore, "Analytical Piecewise Radial Distortion Model for Precision Camera Calibration," *IEE Proceedings—Vision, Image and Signal Processing*, Vol. 153, No. 4, 2006, pp. 468–474.

[107] Hartley, R. I., and A. Zisserman, *Multiple View Geometry in Computer Vision*, Second Edition, Cambridge, U.K.: Cambridge University Press, 2004.

[108] Hartley, R. I., "In Defense of the Eight-Point Algorithm," *IEEE Transactionson Pattern Analysis and Machine Intelligence*, Vol. 19, No. 6, 1997, pp. 580–593.

[109] Nister, D., "An Efficient Solution to the Five-Point Relative Pose Problem," *IEEE Transactions on Pattern Analysis and Machine Intelligence*, Vol. 26, No. 6, 2004, pp. 756–770.

[110] Cattin, P., *Image Restoration: Introduction to Signal and Image Processing*, MIAC, University of Basel, 2012.

[111] Zhou, M., Y. Ji, Y. Ding, J. Ye, S. S. Young, and J. Yu, "Non-Lambertian Surface Shape and Reflectance Reconstruction Using Concentric Multi-Spectral Light Field," *CoRR*, abs/1904.04875, 2019.

[112] Fischler, M. A., and R. C. Bolles, "Random Sample Consensus: A Paradigm for Model Fitting with Applications to Image Analysis andAutomated Cartography," *Communications of the ACM*, Vol. 24, No. 6, June 1981, pp. 381–395.

[113] Fraundorfer, F., and D. Scaramuzza, "Tutorial: Visual Odometry," *IEEE Robot and Automation Magazine*, Vol. 18, No. 4, 2011, pp. 80–92.

[114] Levin, A., and R. Szeliski, "Visual Odometry and Map Correlation," in *Proceedings of the 2004 IEEE Computer Society Conference on Computer Vision and Pattern Recognition, 2004 (CVPR 2004)*, Washington, D.C., 2004, pp. I-I, doi: 10.1109/CVPR.2004.1315088.

[115] Nister, D., O. Naroditsky, and J. Bergen, "Visual Odometry," in *Proceedings of the 2004 IEEE Computer Society Conference on Computer Vision and Pattern Recognition, 2004 (CVPR 2004)*, Vol. 1, 2004, pp. I–I.

[116] Santos, A. C., L. Tarrataca, and J. M. P. Cardoso, "An Analysis of Navigation Algorithms for Smartphones Using J2ME," in *Mobile Wireless Middleware, Operating Systems, and Applications* (J.-M. Bonnin, C. Giannelli, and T. Magedanz, eds.), Berlin: Springer Berlin Heidelberg, 2009, pp. 266–279.

[117] Ruotsalainen, L., "Visual Gyroscope and Odometer for PedestrianIndoor Navigation with a Smartphone," in *25th International Technical Meeting of the Satellite-Division of the Institute of Navigation*, 2012, pp. 2422–2431.

[118] Mur-Artal, R., and J. D. Tardos, "Visual-Inertial Monocular SLAM with Map Reuse," *CoRR*, 2016.

[119] Dong, G., and Z. H. Zhu, "Incremental Visual Servo Control of Robotic Manipulator for Autonomous Capture of Non-Cooperative Target," *Advanced Robotics*, Vol. 30, No. 22, 2016, pp. 1458–1465.

[120] Jirawimut, R., S. Prakoonwit, F. Cecelja, and W. Balachandran, "Visual Odometer for Pedestrian Navigation," *IEEE Transactions on Instrumentation and Measurement*, Vol. 52, No. 4, 2003, pp. 1166–1173.

[121] Coughlan, J., and A. Yuille, "Manhattan World: Compass Direction from a Single Image by Bayesian Interference," in *Proceedings of the Seventh IEEE International Conference on Computer Vision*, IEEE, Vol. 2, 1999, pp. 941–947.

[122] Hough, P. V. C., *Method and Means for Recognizing Complex Patterns*, U.S. Patent 3,069,654, December 18, 1962.

[123] Mutka, A., D. Miklic, I. Draganjac, and S. Bogdan, "ALow Cost Vision Based Localization System Using Fiducial Markers," *IFAC Proceedings Volumes*, Vol. 41, No. 2, 2008, pp. 9528–9533.

[124] Setola, R., G. Ulivi, S. Panziery, and F. Passuci, "A Low Cost Vision Based Localization System for Mobile Robots," in *9th Mediterranean Conference on Control and Automation*, Dubrovnik, Croatia, 2001.

[125] Magnago, V., P. Bevilacqua, L. Palopoli, R. Passerone, D. Fontanelli, and D. Macii, "Optimal Landmark Placement for Indoor Positioning Using Context Information and Multi-Sensor Data," in *2018 IEEE International Instrumentation and Measurement Technology Conference (IPIN MTC)*, 2018, IEEE, pp. 1–6.

[126] Goodfellow, I., Y. Bengio, and A. Courville, *Deep Learning*, MIT Press, 2016, http://www.deeplearningbook.org.

3

Positioning and Navigation Algorithms

This chapter first discusses *static positionings*, which are methods for transforming measurements into positions or position changes. We begin by discussing absolute positioning methods, mainly based on radio signals. Then, we present how the inertial sensor measurements are processed for relative position; that is, change in position with respect to a previous time epoch. Position computation might be considered static from both spatial and temporal viewpoints. The user might be static, for example a cell tower whose precise location we are solving. In this book, as we are addressing navigation, *static positioning* refers to computing the solution for only one time epoch without considering any prior knowledge of the motion. Especially in indoor environments, errors degrade the obtained position and cause solution drift, and therefore we concentrate on error analysis and estimation to improve the results. Next, we look at methods including the temporal domain for a full navigation solution. Such methods are often based on statistical filtering and therefore we discuss Kalman and particle filters. We end the chapter with a look at the future of navigation, which will most likely be based on machine learning for solving various challenges.

3.1 From Measurements to Position: Static Positioning

This section discusses processes required for transforming the various measurements discussed in the previous chapter into observables used in navigation, namely the displacement, range, heading change, angle, and position. For an accurate solution, the errors deteriorating the observables must be modeled and position solved via estimation. Both error modeling and estimation are discussed here.

3.1.1 Ranging

The relation between the user position and a corresponding range measurement to a reference node i can be represented as

$$d_i = \sqrt{(x_u - x_{ti})^2 + (y_u - y_{ti})^2 + (z_u - z_{ti})^2}, \tag{3.1}$$

where (x_u, y_u, z_u) is the user position and (x_{ti}, y_{ti}, z_{ti}) is a known position of a reference node i. With only a single-range measurement, the user can be mapped to the surface of a sphere with center point (x_{ti}, y_{ti}, z_{ti}) and the radius is d_i. Thus, it is not possible to obtain a definite user position, but more measurements are required. A unique position solution can be found by obtaining an additional angle measurement from the same reference node (AOA or AOD), or by obtaining more range measurements from other reference nodes.

In a full 3D positioning scenario, at least four range measurements are generally needed for a unique solution. However, three range measurements already provide two discrete position solutions that can typically be relatively far from each other. Thus, assuming coarse prior information on the user position, for example, based on earlier observations during a tracking process or applying certain floor area limits of a building, it is often possible to eliminate the other position solution and select the correct one. Assuming a 2D positioning scenario—for example, indoor positioning with a known floor—one range measurement less compared to the 3D scenario is required.

Although geometrical analysis provides the fundamental basis for mapping the range measurements to a position, it is not universally applicable when measurement noise is introduced. Whereas in the noiseless scenario the position solution is found at perfect intersection points of spheres (3D) or circles (2D) generated by the range measurements, in noisy scenarios such intersection points do not necessarily exist. In Figure 3.1, noisy range measurements are illustrated for a given user position from three separate reference nodes. From the figure it is clear that there are no good analytic solutions for finding the user position, as the ranging circles do not have a common intersection point. However, using, for example, a least squares method, the position estimate would be obtained somewhere between the circles where the sum of square distances to each circle is minimized.

3.1.2 Angle of Arrival

Considering 3D coordinates, the AOA, and similarly the AOD, can be represented using an angle pair (θ, φ), where θ is an elevation angle and φ is an azimuth angle. The elevation angle θ refers to the angle between the z-axis and the horizontal plane (i.e., the xy-plane), and azimuth angle refers to the angle in horizontal

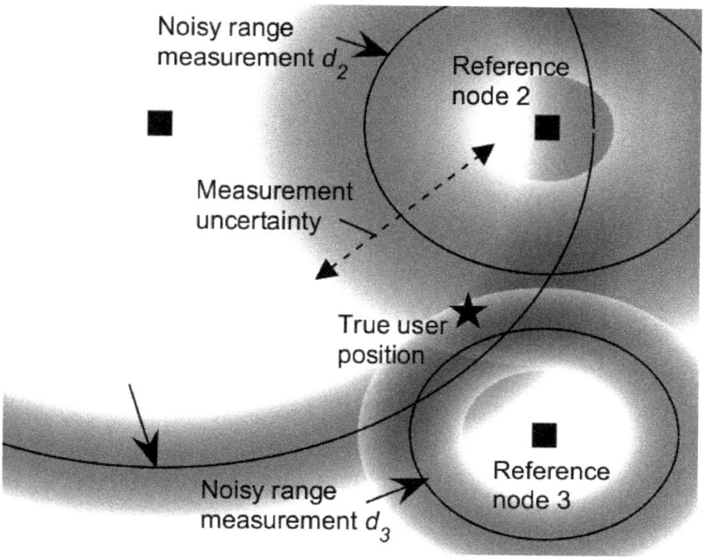

Figure 3.1 Illustration of noisy ranging.

direction (in the *xy*-plane). The definitions of azimuth and elevation angles are related to the device orientation considering roll, pitch, and heading angles, as discussed in Section 2.2.1. Furthermore, when determining the angle, it is always important to consider the used reference coordinate system, as angles can have a meaningful representation in both global and local coordinates. For example, based on a global coordinate system a base station can be said to physically locate north from a user device. However, depending on which direction the user is facing, the angle toward the base stations is varying with respect to a local reference angle at the user. Since there are numerous ways to define and implement device rotations, we omit the rotations from the angle equations. However, for an interested reader, an example of standardized mapping between local and global coordinates and related 3D rotations can be found in [1] for 5G mobile networks.

The relationship between a user position (x_u, y_u, z_u) and the corresponding user angle (θ_i, φ_i) observed at a reference node i can be defined as

$$\varphi_i = \text{arctan}_2(y_u - y_{ti}, x_u - x_{ti}) \text{ and} \tag{3.2}$$

$$\theta_i = \arcsin\left(\frac{z_u - z_{ti}}{d_i}\right), \tag{3.3}$$

where (x_{ti}, y_{ti}, z_{ti}) is the position of the reference node, $\text{arctan}_2(\cdot)$ is a four-quadrant inverse tangent function, and d_i is the 3D distance between the user and reference node as defined in (3.1). Based on the above angle definitions, the zero of the azimuth angle $\varphi_i \in]-\pi, \pi]$ points at the x-axis direction, whereas the zero of the elevation angle $\theta_i \in [-\frac{\pi}{2}, \frac{\pi}{2}]$ points at the horizon. In case it is desired to obtain the angle from the user toward the reference node, only changing the sign of the input parameters in the above functions (e.g., $(x_u - x_{ti}) \to (x_{ti} - x_u)$) is needed. The presented coordinate system with related azimuth and elevation angles is geometrically illustrated in Figure 3.2.

Similar to the ranges discussed in Section 3.1.1, when angle measurements from multiple reference nodes are obtained, it is possible to find a unique mapping between the angle measurements and user position. In practical scenarios with noisy measurements, geometrically derived solutions might not exist, and therefore using alternative methods, such as the least squares, is often preferred. However, when considering both the angle (θ_i, φ_i) and range d_i for a reference node i, it is possible to present the user position as

$$x_u = x_{ti} + d_{2\mathrm{D},i}\cos(\varphi_i) \tag{3.4}$$

$$y_u = y_{ti} + d_{2\mathrm{D},i}\sin(\varphi_i) \tag{3.5}$$

$$z_u = z_{ti} + d_i\sin(\theta_i) \tag{3.6}$$

where

$$\begin{aligned} d_{2\mathrm{D},i} &= \sqrt{d_i^2(1 - \sin^2(\theta_i))} \\ &= \sqrt{(x_u - x_{ti})^2 + (y_u - y_{ti})^2} \end{aligned} \tag{3.7}$$

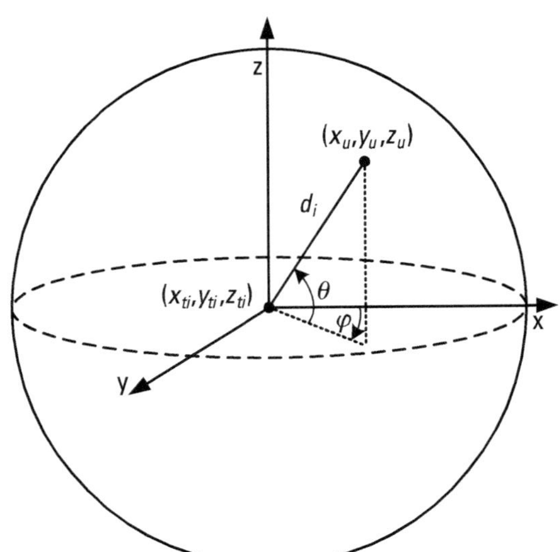

Figure 3.2 Illustration of the coordinate system with related azimuth and elevation angles.

is the 2D distance (i.e., the distance in horizontal plane) between the user and reference node i. When considering a 2D positioning scenario, the z-coordinate can be set to zero, which results in equal 2D and 3D distances.

3.1.3 Strapdown Inertial Navigation

The outputs of inertial sensors (Section 2.2.1) can be converted from specific force and rotation rate measurements to position, velocity, and attitude estimates by means of numerical integration, given appropriate initial conditions. The basic idea is simple: First, the gyro measurement is used to update the estimate of the IMU attitude, conventionally represented in the form of a rotation matrix. Then, the specific force measurement from the accelerometers is converted from the sensor body frame B to the local level frame L, after which the gravity can be added and the resulting acceleration integrated to velocity and position. In mathematical notation, the previous position, velocity, and attitude estimates $\boldsymbol{p}_{t-1}$, $\boldsymbol{v}_{t-1}$, and $\mathrm{R}^L_{B,t-1}$, respectively, are updated with new inertial measurements f^B_t and $\boldsymbol{\omega}^B_t$ as follows:

$$\begin{aligned}
\mathrm{R}^L_{B,t} &= \mathrm{R}^L_B \exp\left(\int_{t-1}^{t} \left[\boldsymbol{\omega}^B_t - \boldsymbol{\omega}^B_{EI}\right]_\times \mathrm{d}\tau\right) \approx \mathrm{R}^L_B\left(\mathrm{I} + \Delta t\left[\boldsymbol{\omega}^B_t - \boldsymbol{\omega}^B_{EI}\right]_\times\right) \\
\boldsymbol{v}^L_t &= \boldsymbol{v}^L_{t-1} + \int_{t-1}^{t} \left(\mathrm{R}^L_{B,\tau} f^B_t + g^L\right) \mathrm{d}\tau \qquad (3.8) \\
\boldsymbol{p}^L_t &= \boldsymbol{p}^L_{t-1} + \int_{t-1}^{t} \boldsymbol{v}^L_\tau \mathrm{d}\tau
\end{aligned}$$

where the exp function is the matrix exponential and Δt is the interval between time epochs $(t - 1)$ and t.

This algorithm is known as the *strapdown* inertial navigation mechanization; the name refers to the fact that the IMU is rigidly attached, strapped down, to the user instead of a dedicated platform that is being mechanically leveled by gimbals, which used to be a common way of implementing inertial navigation before powerful microprocessors became widely available. Equation (3.8) neglects the Coriolis force by assuming the velocity in indoor navigation applications remains so low that the cross product $2\boldsymbol{\omega}^L_{EI} \times \boldsymbol{v}^L$ is negligible. With low-cost MEMS sensors, the measurements are typically so noisy that a simple numerical integration algorithm (e.g., trapezoidal integration) is sufficient; when using high-grade sensors, a numerically a more stable integration method such as fourth-order Runge–Kutta is often applied.

The triple integration sequence in (3.8) is susceptible to error accumulation. By perturbation analysis, considering only additive sensor biases $\delta \boldsymbol{a}$ and $\delta\boldsymbol{\omega}$ as measurement errors, we can derive the following continuous-time system of linear differential equations to model the propagation of position, velocity, and attitude errors δp, δv, and $\delta \psi$, respectively [2]:

$$\frac{\mathrm{d}}{\mathrm{d}t}\begin{bmatrix}\delta p^L\\ \delta v^L\\ \delta \psi^L\\ \delta a^B\\ \delta \omega^B\end{bmatrix} = \begin{bmatrix}\mathrm{O} & \mathrm{I} & & & \\ & \mathrm{O} & [\mathrm{R}_B^L f^B]_\times & \mathrm{R}_B^L & \\ & & \mathrm{O} & & -\mathrm{R}_B^L \\ & & & \mathrm{O} & \\ & & & & \mathrm{O}\end{bmatrix}\begin{bmatrix}\delta p^L\\ \delta v^L\\ \delta \psi^L\\ \delta a^B\\ \delta \omega^B\end{bmatrix} \tag{3.9}$$

where O denotes a 3×3 matrix of zeros, I is the 3×3 identity matrix, and the attitude perturbation $\delta \psi$ is defined by $\widetilde{\mathrm{R}}_B^L = \left(\mathrm{I} - [\delta \psi^L]_\times\right)\mathrm{R}_B^L$. Small attitude offsets expressed in the L-frame can be interpreted as roll, pitch, and heading biases.

3.2 Theoretical Error Analysis

Assuming a set of sensor measurements, user positioning together with estimation of positioning-related parameters, such as ranges or angles, can be practically implemented in countless ways. In order to answer the question of how good the considered estimation method is, it is possible to use theoretical error analysis, which is able to provide a lower bound of the estimator error variance. In addition, when studying or dimensioning a potential performance of a positioning system under certain parameter configurations, for example, related to bandwidth or antenna array sizes, theoretical error analysis provides a good starting point. With theoretical error analysis it is often possible to reduce the need for implementing time-consuming Monte Carlo experiments, including trialing of various estimation methods for understanding the potential estimation accuracies.

3.2.1 Fisher Information and Estimation Error Bounds

Let us consider a parameter vector x, for which we find an unbiased estimator $\hat{x}$. Here, we choose to use the vector definition for the estimated parameter, since it covers estimation scenarios with both a single estimated parameter (i.e., the vector becomes a scalar) and multiple, potentially jointly, estimated parameters. Nonetheless, it can be shown that for any unbiased estimator $\hat{x}$, the CRB provides a theoretical lower bound for the estimator covariance that directly relates to the estimation error [3]. Furthermore, assuming an unbiased estimator $\hat{x}$ for the estimated parameter x, the CRB for the estimator covariance is defined as

$$\mathrm{cov}(\hat{x}_{\mathrm{CRB}}) = (\mathrm{J}(x))^{-1} \leq \mathbb{E}\{(\hat{x} - x)(\hat{x} - x)^T\}, \tag{3.10}$$

where $\mathrm{J}(x)$ is the fisher information matrix (FIM) of the estimated parameter vector x. Moreover, the element in the ith row and jth column of the FIM is defined as [3]

$$[\mathrm{J}(\boldsymbol{x})]_{i,j} = -\mathbb{E}\left\{\frac{\partial^2 \log \mathcal{L}(y|\boldsymbol{x})}{\partial x_i \partial x_j}\right\}, \tag{3.11}$$

where $\mathbb{E}\{\cdot\}$ is the expected value taken with respect to the likelihood function $\mathcal{L}(y|\boldsymbol{x})$, and x_i and x_j are the ith and jth scalar elements of the estimated parameter vector $\boldsymbol{x}$. Fisher information can be thought of as a measure of the amount of information on the unknown parameter $\boldsymbol{x}$ carried by an observed measurement y. Furthermore, the connection between the observed measurement and the unknown parameter is addressed via the likelihood function that characterizes the probability of observing y given $\boldsymbol{x}$.

3.2.2 Error Bound for Propagation Time Estimation

Ranging via estimation of propagation time, also referred to as TOA or TOF, is one of the most significant measurement sources in enabling high positioning accuracy. For example, compared to angle estimation, propagation time estimation does not require extensively large antenna arrays, which is often unfeasible. In addition, unlike with angle estimation, ranging-based positioning accuracy does not deteriorate as a function of distance to the reference node. However, time-measurement-based solutions require careful management of all clocks participating in the measurement process. Although with certain positioning methods, such as TDOA, a clock bias between the user and network can be neglected, accurate synchronization inside the network is still required. In the following discussion we focus on the received signal related propagation time estimation and neglect the possible clock biases. Nonetheless, clock errors at user device and different reference nodes can be separately estimated and properly handled by obtaining conventional propagation time measurements from several reference nodes [4].

Assuming a transmitted, generally complex-valued, continuous signal $s(t)$, the nth received signal sample at the receiver can be given as

$$r(n) = s(nT_s - \tau) + w(n), \tag{3.12}$$

where T_s is the sample interval, which is inversely proportional to the sampling frequency $F_s = \frac{1}{T_s}$. The desired propagation time parameter is denoted as τ, and $w_m \sim \mathcal{N}\left(0,\, \sigma_w^2\right)$ represents additive white Gaussian noise with variance σ_w^2. Now, it can be shown that the CRB for estimating the propagation time τ is given as [5]

$$\mathrm{var}(\hat{\tau}) \geq \frac{\sigma_w^2}{2\sum_{n=-\infty}^{\infty}\left|\frac{\mathrm{d}}{\mathrm{d}\tau}s(nT_s - \tau)\right|^2}. \tag{3.13}$$

Besides the obvious relation between the CRB and noise variance, and consequently the signal-to-noise ratio (SNR), the CRB is affected by the derivative of the transmitted signal. Here, the SNR represents the ratio of the useful signal power to the noise power, and is an essential indicator of received signal quality. Based on the derivative term in the above CRB, it seems that by increasing the signal variations, the estimator error can be reduced. As, generally, time-variations of a signal are related to the signal spectrum, it is reasonable to study the CRB in frequency domain using a Fourier transform. It turns out that using a frequency domain representation, the CRB for propagation time estimation is

$$\mathrm{var}(\hat{\tau}) \geq \frac{\sigma_w^2 T_s}{8\pi^2 \int_{-\frac{F_s}{2}}^{+\frac{F_s}{2}} f^2 |S(f)|^2 \mathrm{d}f}. \tag{3.14}$$

where $S(f)$ is the spectrum (i.e., a Fourier transformation) of the transmitted signal $s(t)$, and $F_s = \frac{1}{T_s}$ is the sampling frequency. Inside the integral the frequency term $f \in \left[-\frac{F_s}{2}, \frac{F_s}{2}\right]$ is squared, which indicates that the estimation error can be reduced by allocating transmitted signal energy toward the band edges. Since high frequencies are associated with high signal variations in time, this observation is in line with the CRB result in (3.13), where high signal variations were seen to improve the estimation accuracy.

Many modern and commonly used wireless communications systems use multicarrier modulation, especially the OFDM, as their main modulation scheme. Such systems include, for example, 4G LTE and 5G NR mobile networks, and WLAN 801.11 standard. Assuming an OFDM signal within total of K subcarriers $k = \lfloor -\frac{K-1}{2} \rfloor, \ldots, \lfloor \frac{K-1}{2} \rfloor$, the CRB for propagation time estimation is given as

$$\mathrm{var}(\hat{\tau}) \geq \frac{\sigma_w^2}{8\pi^2 \Delta_f^2 \sum_{k=\lfloor -\frac{K-1}{2} \rfloor}^{\lfloor \frac{K-1}{2} \rfloor} k^2 |S_k|^2}, \tag{3.15}$$

where Δ_f is the subcarrier spacing, and S_k is the transmitted symbol at the kth subcarrier. Similar to (3.14), allocating more power towards the edge subcarriers reduces the estimation error of the propagation time.

In Figure 3.3, the CRB for the OFDM-based propagation time estimation is illustrated for different SNRs as a function of signal bandwidth $K \cdot \Delta_f$. The CRBs shown assume equal power distribution over all subcarriers, and the used subcarrier spacing is $\Delta_f = 60$ kHz. Moreover, instead of indicating the CRB as a variance of the propagation time, the CRB is given as a standard deviation of the estimated range that is connected to the standard deviation of propagation time as $\mathrm{std}(\hat{d}) = c \cdot \mathrm{std}(\hat{\tau})$, where c is the speed of light. Besides the obvious accuracy increment by elevating the SNR, it is clear that the accuracy of propagation time estimates can be improved by increasing the bandwidth.

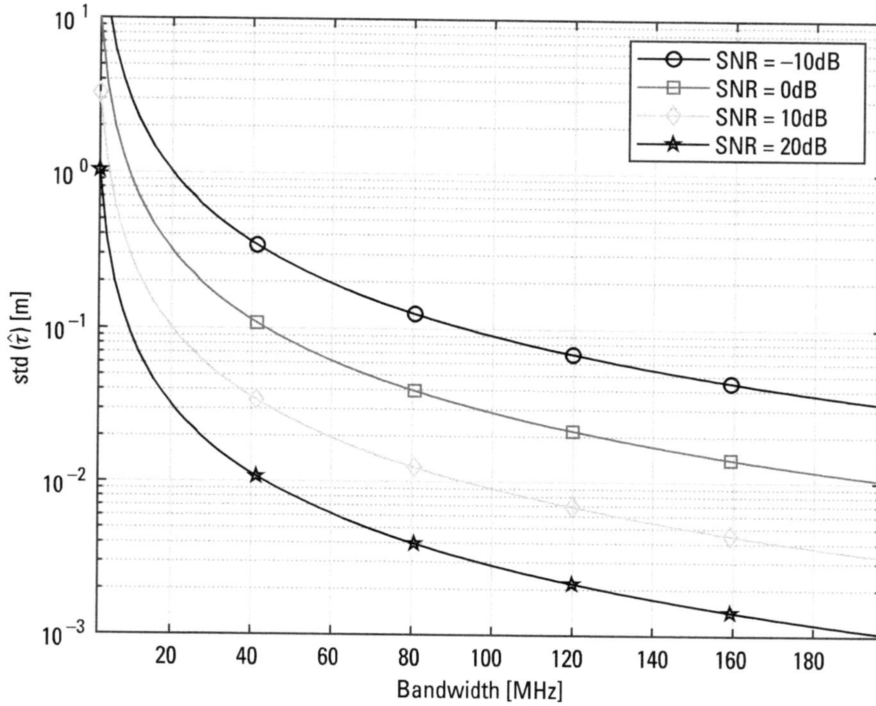

Figure 3.3 Illustration of error bounds for OFDM-based propagation time estimation as a function of bandwidth. The CRB is shown for four different SNRs.

3.2.3 Error Bound for Angle Estimation

Estimation of AOA or AOD is generally achievable, if the observed signal is somehow dependent on the considered angle parameter. In practice, it means using directive antennas or antenna arrays that can be steered towards different directions mechanically and/or electronically. In the following discussion, we first consider antenna arrays using an assumption of uniform linear array (ULA), and then move to generic sectorized antennas using simple sector-wise power measurements that can be conveniently applied for both mechanically and electronically steered antennas.

In the case of antenna arrays, the antenna beam can be electrically steered toward a desired direction by exploiting a known antenna element geometry. As an example, in Figure 3.4, a ULA with four antenna elements is illustrated. There, a wavefront is arriving from angle φ, and due to the given array structure, the wavefront has a different propagation distance to each antenna element. Based on the ULA geometry, the difference in the propagation distance between two consecutive antenna elements can be found as $\Delta d = d_{\text{ant}} \sin(\varphi)$, where d_{ant} is the antenna separation distance. Now, by utilizing the antenna element geometry

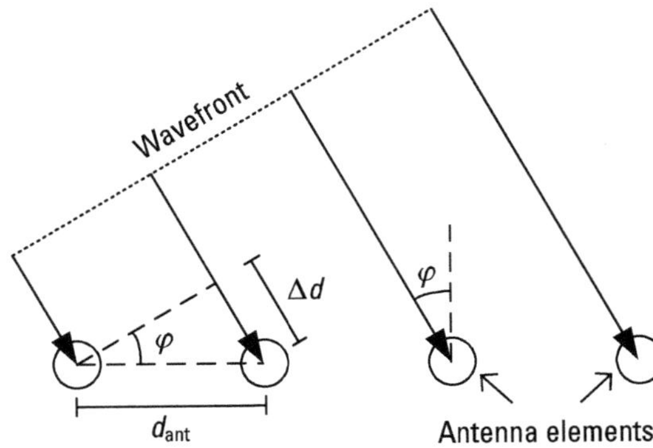

Figure 3.4 Illustration of ULA with an incoming wavefront from angle φ.

shown in Figure 3.4, the received signal at the mth antenna element can be written as

$$r_m = Ae^{j\left(\psi - \left(m - \frac{M-1}{2}\right)\frac{2\pi d_{\text{ant}}}{\lambda}\sin(\varphi)\right)} + w_m, \tag{3.16}$$

where A and ψ are constant amplitude and phase parameters related to the received signal characteristics. Due to independence from antenna element index m, A and ψ are equal between all elements and do not directly contribute to the angle estimation. In addition, M is the number of antenna elements, λ is the signal wavelength, φ is the AOA, and w_m denotes the element-wise noise term. It should be noted that the above modeling as well as the following CRB analysis is equivalently usable for the AOD by changing the signal direction.

Assuming white Gaussian noise as $w_m \sim \mathcal{N}\left(0,\, \sigma_w^2\right)$, where σ_w^2 is the noise variance, the CRB for the ULA-based AOA can be given as [5]

$$\text{var}(\hat{\varphi}) \geq \frac{6\lambda^2}{\gamma M\left(M^2 - 1\right)\left(2\pi d_{\text{ant}}\cos(\varphi)\right)^2}, \tag{3.17}$$

where $\gamma = \frac{A^2}{\sigma_w^2}$ is the SNR of the received signal. Based on the above, the AOA accuracy is proportional to M^{-3}, and thus can be considerably improved by increasing the number of used antenna elements M. One particularly interesting feature of the AOA estimation CRB is that the accuracy is dependent on the value of the AOA itself via the cosine term in the nominator. The accuracy is the highest when $\varphi = 0$, meaning that the AOA is directly ahead of the array, and the accuracy reduces when moving toward the sides $\varphi \pm 90°$, where the error variance approaches to infinity. This is due to the fact that for the ULA the array factor and related angular resolution changes as a function of AOA. The same array factor can also be affected by the wavelength λ and the antenna separation distance d_{ant}. However, in many wireless communications systems the antenna separation distance is limited to $d_{\text{ant}} \leq \lambda/2$ in order to avoid multiple maxima in array beamforming.

In Figure 3.5, the CRB for the ULA-based AOA estimation is depicted as a function of number of antenna elements, while assuming $d_{\text{ant}} = \lambda/2$. The CRB is evaluated for three different SNRs and two different AOAs. Based on the figure, besides increasing the SNR, it is evident that increasing the number of antenna elements improves the estimation accuracy. The AOA itself depends on the user position and often cannot be influenced by the estimator design. However, it is clearly indicating that for providing a good estimation coverage in a positioning system, careful design of array positions and their orientations is needed.

In many practical scenarios, utilization of antenna arrays can be limited due to several implementation-related issues, such as unknown phase offsets between the elements. In addition, obtaining antenna-element-wise samples of the received signal requires digital array processing with separate transmission chains behind each antenna element. This necessitates, for example, element-wise amplifiers, analog-to-digital converters, and related digital signal processing, which increases the complexity. Thus, in many use cases, analog beamforming is used, where the received signals of all antenna elements are summed together

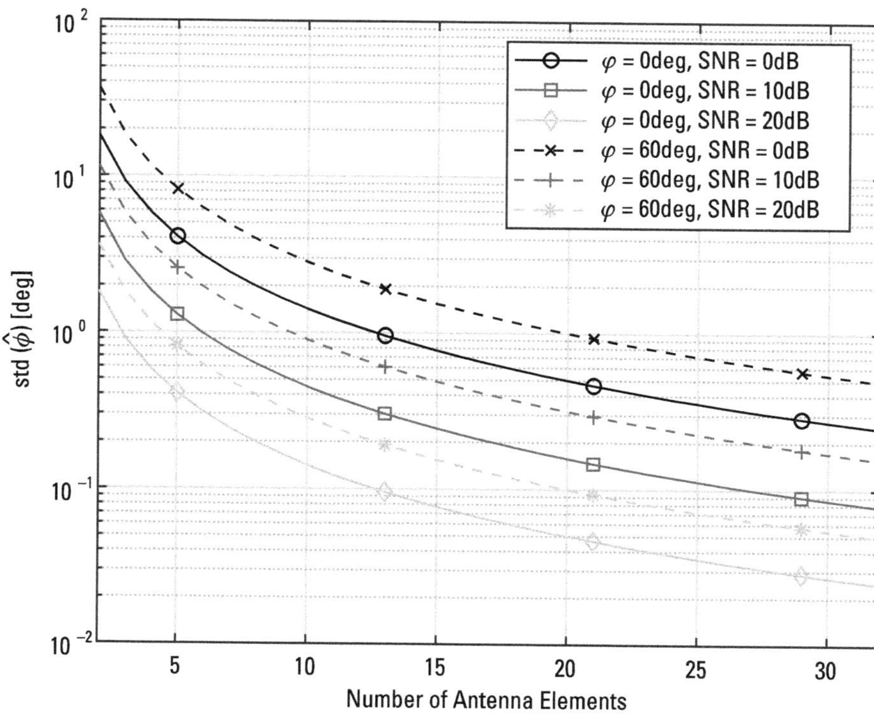

Figure 3.5 Illustration of error bounds for ULA-based angle estimation as a function of number of antenna elements. The CRB is shown for three different SNRs and two different AOAs.

before sampling. In this case, the received (single-stream) digital signal becomes $r = \sum_m r_m e^{j\zeta_m}$, where ζ_m is the phase-shift applied for the mth antenna element. Since theoretically only a single beam direction can be covered with one combination of phase shifts, observing unknown angles requires searching with different beams over all target angles. This procedure is called beam sweeping, and it is one fundamental ingredient, for example, in 5G NR systems.

If all phase-shifter information is available and the channel remains constant during a beam sweeping procedure, it is possible to use similar angle estimation techniques as with fully digital ULA. However, with analog beamformers (i.e., observing only the phase-shifted sum of the element-wise signals) it is more common to only consider beam-wise power measurements for angle estimation. With certain limitations, this can be compared to a mechanically rotating directive antenna, which obtains signal power measurements over a set of scanned directions. In order to study these types of measurements for angle estimation, a concept of sectorized antenna can be exploited [6]. By using a sectorized antenna, the nth received signal sample from the mth sector is given as

$$r_m(n) = \Psi(\varphi - \upsilon_m)s(n) + w_m(n), \tag{3.18}$$

where $\Psi(\cdot)$ is the antenna pattern function represented in amplitude domain, υ_m is the direction angle of the mth sector, $s(n)$ is the received signal including possible propagation losses, and $w(n)$ is the noise term. A simple example of the antenna radiation pattern can be presented using a Gaussian-like pattern as

$$\Psi(\varphi) = e^{-\frac{(\mathcal{M}(\varphi))^2}{\beta}}, \tag{3.19}$$

where $\mathcal{M}(\varphi) = \mathrm{mod}_{2\pi}(\varphi + \pi) - \pi$ takes into account the cyclic nature of the angle and β is a parameter to control the sector width. An illustration of Gaussian-like sector (or beam) patterns considering 12 uniformly separated sectors over the full circle, is shown in Figure 3.6. Here, the sector width has been selected so that the maximum amplitude is suppressed by a factor of 10 at the center of a neighboring sector.

A power measurement from the mth sector can be obtained as $P_m = \frac{1}{N}\sum_n |r_m(n)|^2$. Based on the signal model in (3.18), and assuming white Gaussian noise as $w_m(n) \sim \mathcal{N}\left(0, \sigma_w^2\right)$ and a known path loss, the CRB for the AOA using sector (or beam) associated power measurements can be given as [6]

$$\begin{aligned} \mathrm{var}(\hat{\varphi}) &\geq \frac{1}{\mathcal{J}(\varphi)}, \text{ where} \\ \mathcal{J}(\varphi) &= (N+2)\sum_m \frac{\tilde{\rho}_m^2(\varphi)}{\left(1 + \frac{1}{\Psi^2(\varphi - \upsilon_m)\gamma}\right)^2} \end{aligned} \tag{3.20}$$

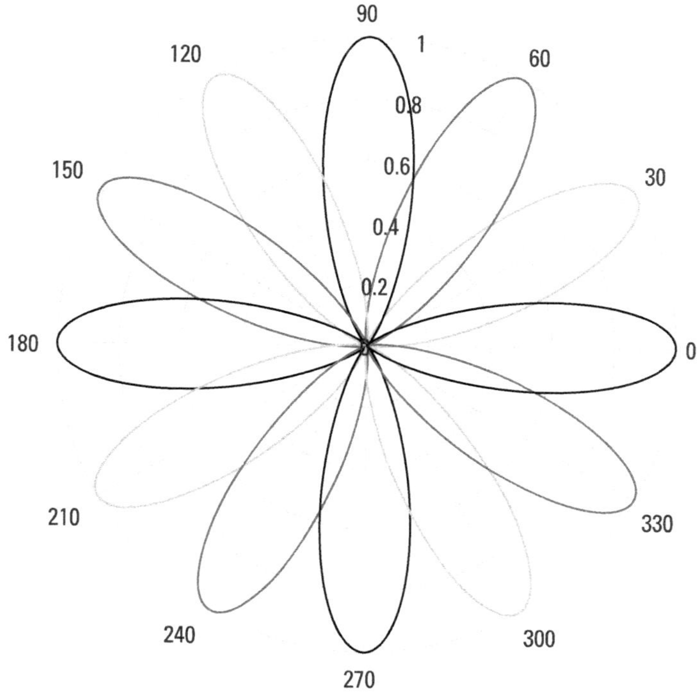

Figure 3.6 Illustration of Gaussian-like beam pattern amplitudes considering 12 uniformly distributed beam angles over 360° degrees. Each beam pattern is normalized to have a maximum value of 1 at the main beam direction, and the beam amplitude at the neighboring beam direction is set to 0.1, which results in 20-dB power suppression for the neighboring beam.

denotes the Fisher information. Furthermore, N is the number of obtained samples, γ is the SNR of the received signal, and

$$\tilde{\rho}_m(\varphi) = \Psi^{-2}(\varphi - \upsilon_m)\frac{d\Psi^2(\varphi - \upsilon_m)}{d\varphi} \tag{3.21}$$

is the ratio between the sector power pattern and its derivative with respect to the angle φ. For example, considering the Gaussian-like antenna pattern described in (3.19), $\tilde{\rho}_m(\varphi) = -4\frac{\mathcal{M}(\varphi-\upsilon_m)}{\beta}$. Besides angle estimation, it is also possible to find the CRB for a joint estimator of angle and pathloss, as shown in [6].

In Figure 3.7 the CRB for the sector (or beam) based AOA estimation using sector-wise power measurements is illustrated as a function of the number of sectors/beams. The sectors are defined according to Gaussian patterns, similar to (3.19), and are uniformly distributed over all directions. Furthermore, each sector has a 20-dB power reduction toward the direction of a neighboring sector, and the number of samples for each power measurements is $N = 10$. Since

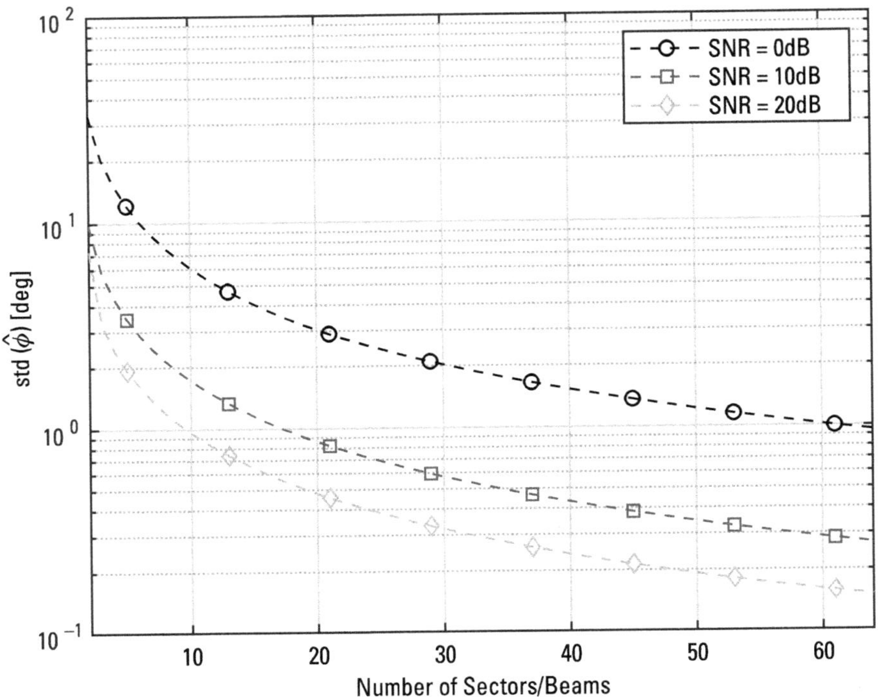

Figure 3.7 Illustration of error bounds for angle estimation using sector- (or beam-) wise power measurements as the number of antenna elements. The antenna patterns are assumed Gaussian-like with 20-dB power reduction for the neighboring sector. The CRB is shown for three different SNRs.

the CRB is dependent on the AOA, the CRB curves shown are a result of an average of 1,000 individual CRBs from uniformly distributed AOAs between two neighboring sectors. It should be noted that due to the assumed symmetry of the sectors, the CRBs between a pair of sectors can be generalized to any other pair of sectors. From the figure, it can be clearly seen that increasing the number of sectors improves the estimation accuracy. This is intuitive, as including more sectors implies increased angular resolution.

3.2.4 Position Error Bound

Positioning-related measurements, such as range and angle measurements, can be associated with the user position via straightforward geometric operations, as shown in Sections 3.1.1 and 3.1.2. This enables converting the obtained Fisher information from the angle and propagation time estimation into a position error bound (PEB) [3, 7, 8], which provides a CRB for the positioning error. In order to define the bound, let's define a measurement vector $\boldsymbol{\theta}$, which contains all the measured parameters used in the positioning solution, such as range and angle measurements for a specific set of reference nodes. Furthermore, for each measurement there is a specific mapping function that associates the measurement with a user position, as given in (3.1) for range measurements and (3.2) and (3.3) for angle measurements. Considering the above, at a user position $\boldsymbol{x} = [x_u \;\; y_u \;\; z_u]^T$, the PEB is given as

$$\text{PEB} = \sqrt{\text{trace}((\text{H}(\boldsymbol{x})^T \text{J}(\boldsymbol{\theta})\text{H}(\boldsymbol{x}))^{-1})}, \tag{3.22}$$

where $\text{H}(\boldsymbol{x})$ is the Jacobian matrix of the measurement model with respect to the position $\boldsymbol{x}$, and $\text{J}(\boldsymbol{\theta})$ is the FIM for the estimated parameter vector $\boldsymbol{\theta}$, such as ranges and angles.

To illustrate an example of the required Jacobian matrix for the above PEB evaluation, let's consider a set of measurements for a single reference node given as a vector $[\tau \;\; \varphi \;\; \theta]^T$, where the individual elements represent measurements of propagation time, azimuth angle and elevation angle, in respective order. Consequently, the measurement model, defining the mapping between the measurements and the desired user position, is given as

$$\boldsymbol{h}(\boldsymbol{x}) = \begin{bmatrix} \tau \\ \varphi \\ \theta \end{bmatrix} = \begin{bmatrix} \frac{\sqrt{(x_u - x_{ti})^2 + (y_u - y_{ti})^2 + (z_u - z_{ti})^2}}{c} \\ \text{atan2}(y_u - y_{ti}, x_u - x_{ti}) \\ \arcsin(\frac{z_u - z_{ti}}{d_i}) \end{bmatrix}, \tag{3.23}$$

where c is the speed of light. The Jacobian matrix $\text{H}(\boldsymbol{x}) \in \mathbb{R}^{3\times3}$ can then be obtained as

$$\text{H}(\boldsymbol{x}) = \begin{bmatrix} \frac{\partial[\boldsymbol{h}(\boldsymbol{x})]_1}{\partial x_u} & \frac{\partial[\boldsymbol{h}(\boldsymbol{x})]_1}{\partial y_u} & \frac{\partial[\boldsymbol{h}(\boldsymbol{x})]_1}{\partial z_u} \\ \frac{\partial[\boldsymbol{h}(\boldsymbol{x})]_2}{\partial x_u} & \frac{\partial[\boldsymbol{h}(\boldsymbol{x})]_2}{\partial y_u} & \frac{\partial[\boldsymbol{h}(\boldsymbol{x})]_2}{\partial z_u} \\ \frac{\partial[\boldsymbol{h}(\boldsymbol{x})]_3}{\partial x_u} & \frac{\partial[\boldsymbol{h}(\boldsymbol{x})]_3}{\partial y_u} & \frac{\partial[\boldsymbol{h}(\boldsymbol{x})]_3}{\partial z_u} \end{bmatrix} \tag{3.24}$$

$$= \begin{bmatrix} \frac{x_u - x_{ti}}{cd_i} & \frac{y_u - y_{ti}}{cd_i} & \frac{z_u - z_{ti}}{cd_i} \\ \frac{y_{ti} - y_u}{d_{2\text{D},i}} & \frac{x_u - x_{ti}}{d_{2\text{D},i}} & 0 \\ \frac{(x_{ti} - x_u)(z_u - z_{ti})}{d_i^2 d_{2\text{D},i}} & \frac{(y_{ti} - y_u)(z_u - z_{ti})}{d_i^2 d_{2\text{D},i}} & \frac{d_{2\text{D},i}}{d_i^2} \end{bmatrix}, \tag{3.25}$$

where $[\boldsymbol{h}(\boldsymbol{x})]_i$ denotes the ith element of vector $\boldsymbol{h}(\boldsymbol{x})$. In addition, d_i and $d_{2\text{D},i}$ are the 3D distance and 2D distance between the user and reference node i, as defined in (3.1) and (3.7), respectively. The number of columns in the Jacobian matrix is determined by the number of estimated parameters, which here include the user x-coordinate, y-coordinate, and z-coordinate. If desired, other estimated parameters, such as unknown clock biases, antenna orientations, or other positioning-related parameters, can also be involved. However, increasing the number of estimated parameters typically brings about the need for more measurements. For example, instead of obtaining the propagation time, azimuth angle, and elevation angle for a single-reference node, one could consider multiple reference nodes, and thus multiply the number of rows in the Jacobian matrix.

In the following, we illustrate a 2D PEB for a specific geographical area with three reference nodes using estimates of propagation time and azimuth angle. In order to neglect the effect of user orientation, the angles are estimated at the fixed reference nodes, where the orientation is assumed to be known. The considered estimates (or measurements) of propagation times and azimuth angles are assumed to be uncorrelated, and all the estimates are assumed to achieve the CRBs according to (3.15), and (3.17) or (3.20) depending on the angle estimation scheme used. At each location the SNR in logarithmic scale is defined as $\gamma[dB] = P_R[\text{dBm}] - P_N[\text{dBm}]$, where $P_R[\text{dBm}] = P_T[\text{dBm}] - L[\text{dB}]$ is the received signal power, and $P_N[\text{dBm}] = -174 + 10\log_{10}(B) + 6$ is the thermal noise power at room temperature with a bandwidth of B and receiver noise figure of 6 dB. Moreover, $P_T[\text{dBm}] = 27\,\text{dBm} \approx 500\,\text{mW}$ is the transmitted power and $L[\text{dB}] = 60 + 30\log_{10}(d)$ is the path loss, which is determined roughly based on the path loss models for indoor office and indoor factory by the 3GPP for the 5G NR systems in [1]. The assumed carrier frequency and the effective bandwidth are fixed to 30 GHz and 36 MHz, respectively, while the modulation used is OFDM with 600 active subcarriers and subcarrier spacing of $\Delta_f = 60\text{kHz}$. When performing the evaluations according to the above definitions, the median SNR for each reference node is noted to vary between 9–11 dB.

In Figure 3.8, the PEB for the geographical area considered is illustrated for the propagation time based measurements with reference nodes positions given as $(0, 25)$, $(0, 0)$, and $(50, 0)$. In addition, the reference node array orientations,

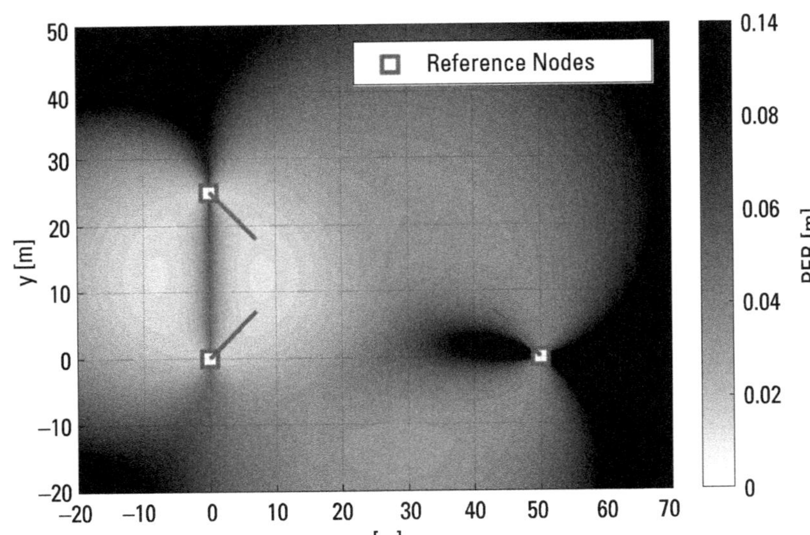

Figure 3.8 PEB based on propagation time estimation (ranging) with three reference nodes.

pointing at South-East, North-East, and North-West in respective order, are shown in the figure. However, the array orientations play a role only with angle-based measurements and can be neglected with the time-based measurements. The shape of the PEB illustration depends on the SNR of each reference node and most importantly on the geometrical relations of the time measurements as described by the Jacobian matrix in (3.22). In the figure, around the two reference nodes at the left, the positioning error is at a lower level compared to the surroundings of the reference node at the right. This is because on the left-hand side part of the area the proximity of two reference nodes enables two measurements with good SNR, whereas on the right-hand side part of the area, there is only a single reference node nearby. However, even with a good SNR, it is not always possible to achieve good accuracy due to bad positioning geometry. In Figure 3.8, this can be seen as the dark area toward the left from the right-hand side reference node, and between the two reference nodes at the left. From a geometric perspective, the intersection points of the noisy time-estimation-based ranging circles can be seen as indefinite in these locations.

The corresponding PEB illustrations when using azimuth angle measurements are shown in Figure 3.9 for ULA-based estimation, and in Figure 3.10 for sector-based estimation with sector-wise power measurements. Since the results in both figures rely on the same azimuth angle measurements, the geometrical mappings between the angles and user position, described by the Jacobian matrix in (3.22), are identical. Because the SNRs are also equal, the only difference lies in the angle estimation CRBs used. For the ULA, we assume 16 antenna elements with antenna separation $d_{\text{ant}} = \lambda/2$. Moreover, for the sector-based

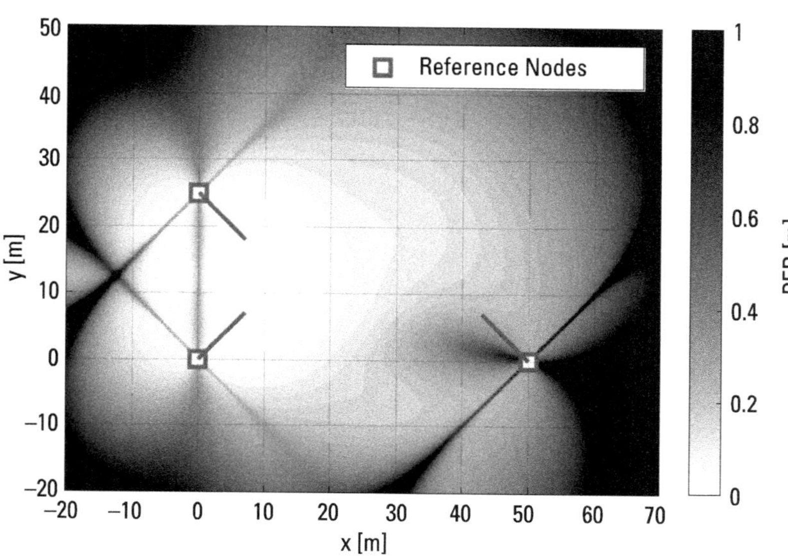

Figure 3.9 PEB based on ULA-based angle estimation with three reference nodes.

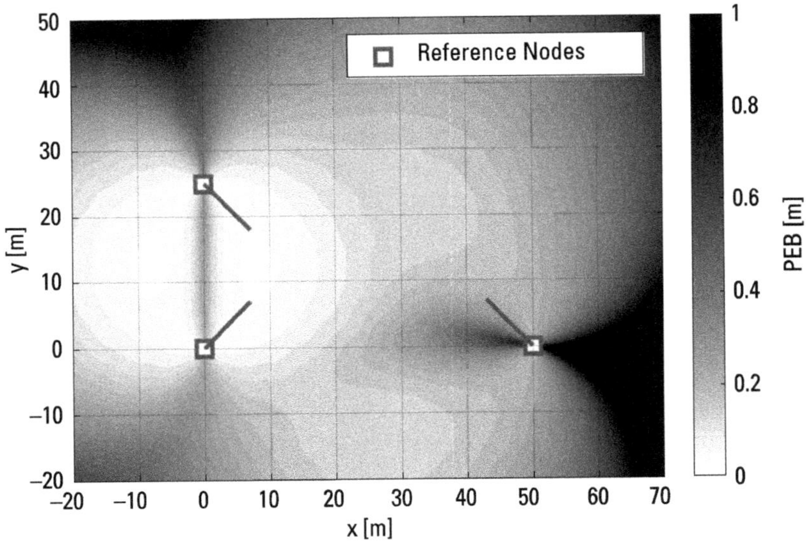

Figure 3.10 PEB based on angle estimation using sector- (or beam-) wise power measurements with three reference nodes.

approach, we assume Gaussian-like power patterns with 6-dB power suppression toward the neighboring sector direction. There are 32 sectors that are uniformly distributed around the full circle, and the power measurement of each sector is based on observations of $N = 100$ samples. Probably the most obvious difference between the ULA-based approach in Figure 3.9 and the sector-based approach in Figure 3.10 is related to asymmetric estimation accuracy of the ULA. As seen earlier in (3.17), the angle estimation accuracy of a ULA is dependent on the angle itself, and the accuracy decreases when the angle is deviated from zero. This is seen in Figure 3.9 as the darker stripes at $\varphi \pm 90^\circ$ directions from the orientation pointer of each reference node.

The PEB for the combined propagation time and angle measurements is illustrated in Figure 3.11. There, the angle measurements are based on the sector-based approach, and all the parameters are identical with the PEB results shown earlier independently for the propagation time based approach and angle-based approach. Although the SNR distribution is unchanged, the positioning accuracy with the combined time- and angle-based measurements seems to be significantly improved compared to approaches using individual time measurements or angle measurements. This is due to improved measurement geometry, which is further illustrated in the following example of corridor-like positioning scenario.

The complementary relationship between angle and range measurements can be conveniently illustrated in a corridor-like environment, where two reference nodes are located at both ends of the corridor. We consider a corridor of 35m in length covering x-coordinates $x \; = \in [-5, 30]$, and 6m in width

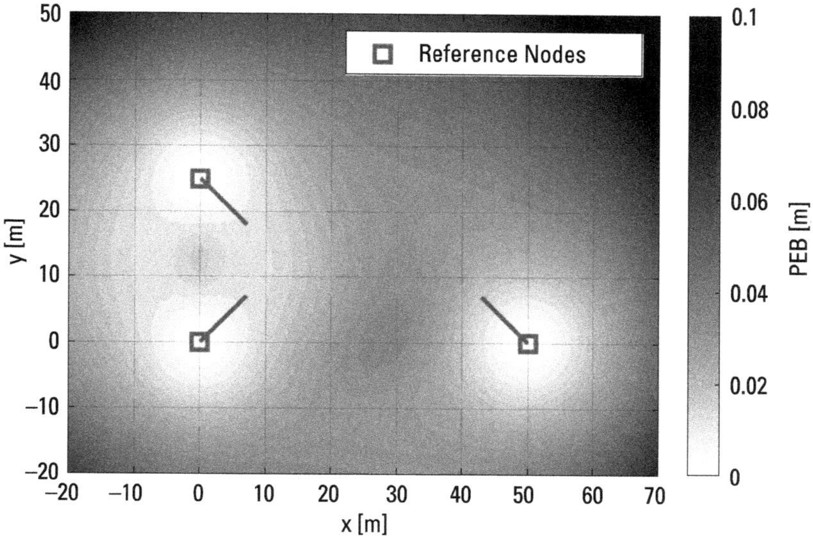

Figure 3.11 PEB based on joint propagation time and angle estimation with three reference nodes.

covering y-coordinates $y =\in [-3, 3]$. There are two reference nodes at positions $(0, -1)$ and $(25, 1)$, as shown in Figure 3.12, which illustrates the PEBs for the range-based (propagation time) estimation, angle-based estimation using the sector approach, and the combined range and angle-based estimation. All parameters are identical with the corresponding PEBs shown earlier, except that the number of samples N in sector-based angle estimation has been tripled for better visualization. The PEBs using only range measurements or only angle measurements show rather poor positioning performance, which is due to a challenging measurement geometry. However, when considering the range and angle measurements jointly, the PEB is significantly improved. How can two poor positioning results, that is, individual range- and angle-based results, together provide such a good outcome?

The solution to the above question lies in studying the estimation accuracy of the x-coordinate and y-coordinate separately. In Figure 3.13, the PEB is shown separately for range- and angle-based approaches considering only estimation of either the x-coordinate or y-coordinate, while assuming the other to be known. From the geometric perspective, an imagined intersection point of two noisy ranging circles is quite definite in the x-direction but indefinite in the y-direction. Exactly the opposite occurs with angle measurements, where an imagined interaction point of two noisy angle-pointers from the reference nodes is quite definite in the y-direction, but indefinite in x-direction. Thus, the range and angle measurements can undoubtedly complement each other from the geometric perspective, and therefore, it is extremely beneficial to consider these aspects in

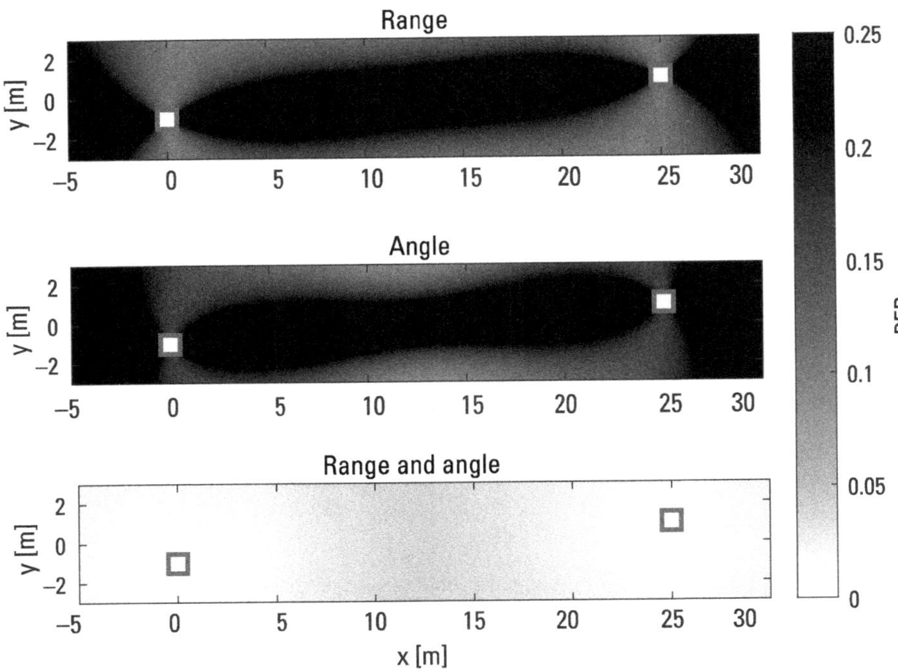

Figure 3.12 Position error bound based on range estimation (top), angle estimation (middle), and joint range and angle estimation (bottom) with three reference nodes in a corridor-like environment.

positioning system design. Depending on the use case, other sensors can also be used to alleviate the geometry-related performance gaps related to range and angle measurements.

An example MATLAB code is provided for illustrating PEB using only angle measurements, only range measurements, or both range and angle measurements for desired base station locations. By default, the code defines the range and angle estimation accuracy based on CRBs according to chosen simulation parameters. However, the code can be easily modified for other estimation accuracies, as well as for different system and base station configurations.

3.3 Least-Squares Estimation

Least-squares (LS) estimation is a method for inferring $\mathbf{x}$, a vector of unknown parameters using $\mathbf{y}$, a set of noisy observations linearly related to it. The relationship between $\mathbf{x}$ and $\mathbf{y}$ is described with

$$\mathbf{y} = \mathbf{A}\mathbf{x} + \mathbf{n}. \tag{3.26}$$

If the number M of unknowns $\mathbf{x} = [x_1, \ldots x_M]^T$ is smaller than the number N of observations $\mathbf{y} = [y_1, \ldots y_N]^T$ the equation is called overdetermined, and if

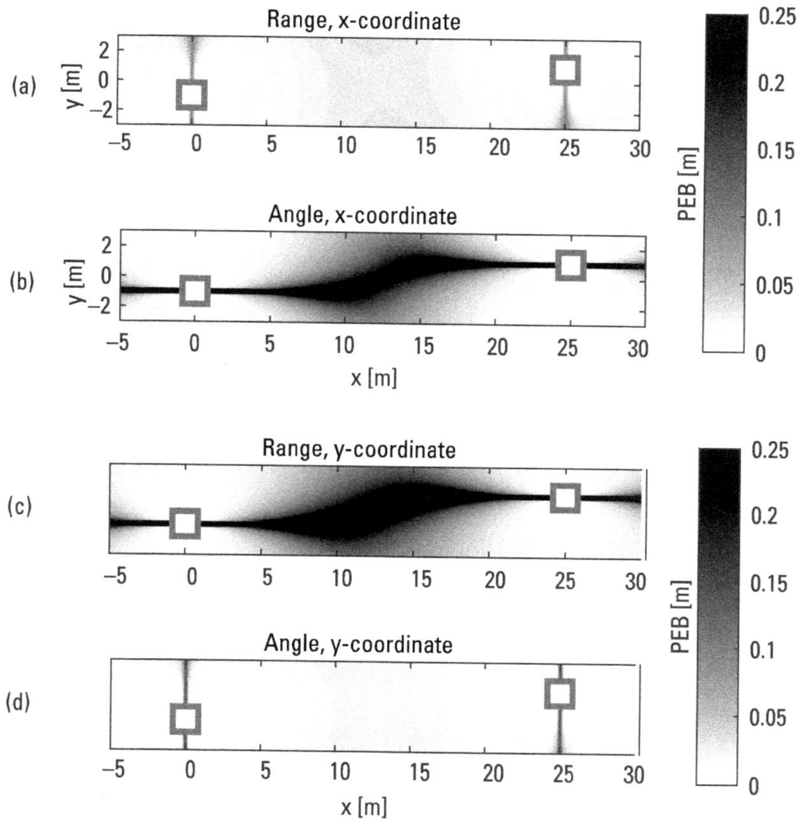

Figure 3.13 PEB considering the estimation of the x-coordinate (a, b) and y-coordinate (c, d) independently, while assuming the other to be known. The PEB is shown separately for range estimation (a, c) and angle estimation (b, d) with two reference nodes in a corridor-like environment.

it is larger the equation is underdetermined. Matrix $\mathbf{A}$, the design matrix, is of size $N \times M$ and contains parameters defining the relationship. The difference between the real values of $\mathbf{x}$ and the ones predicted using (3.26) is denoted with an N-dimensional vector $\mathbf{n}$, the residual. The task then becomes to minimize the residual; that is, find the least value for the sum of squared errors. In other words, the task is to find a *maximum likelihood estimate* (mle) $\hat{\mathbf{x}}$ for $\mathbf{x}$

$$\hat{\mathbf{x}} = \arg\min_{x} \|Ax - y\|^2. \tag{3.27}$$

To get a correct solution for the mle, the observation errors $\mathbf{n}$ must be identically Gaussian distributed with zero mean and statistically independent. After differentiating $||\mathbf{A}\hat{\mathbf{x}} - \mathbf{y}||^2$ with respect to $\hat{\mathbf{x}}$ and finding the minimum at zero, assuming that $\mathbf{A}^T\mathbf{A}$ is not singular, we get a solution for (3.27), the least-squares estimate

$$\hat{\mathbf{x}} = (\mathbf{A}^T\mathbf{A})^{-1}\mathbf{A}^T\mathbf{y}. \tag{3.28}$$

The observation noise, n, is assumed to follow Gaussian distribution with zero mean and covariance matrix $\mathbf{R}$. As the noise distorts the system of linear equations (3.26) will be inconsistent and the inverse of observation covariance matrix $\mathbf{R}^{-1}$ needs to be included for weighting the observations. The weighted least-squares estimate is defined as

$$\hat{\mathbf{x}} = (\mathbf{A}^T\mathbf{R}^{-1}\mathbf{A})^{-1}\mathbf{A}^T\mathbf{R}^{-1}\mathbf{y} \tag{3.29}$$

and its covariance matrix $\mathbf{P}$ is obtained as

$$\mathbf{P} = \left(\mathbf{A}^T\mathbf{R}^{-1}\mathbf{A}\right)^{-1}. \tag{3.30}$$

The LS-equation is usually solved by numerical methods, such as SVD [9]. Quite often the set of positioning equations is not linear and therefore needs to be linearized. Linearization may be done, if an approximation of the user position is available, by expanding the equations in a Taylor series about the approximate position [10]. The following section presents the most popular method for solving nonlinear least-squares problems, Gauss–Newton.

3.3.1 Gauss–Newton Method for Nonlinear Least Squares

The Gauss–Newton algorithm is an iterative method for nonlinear least-squares estimation [11]. The algorithm computes a sequence of linear least-squares approximations to the nonlinear problem. The Gauss-Newton algorithm does not require the evaluation of second-order derivatives in the Hessian of the objective function and has therefore smaller computational cost compared to traditional methods.

The general process of nonlinear least squares is to minimize $|f(x)|^2$, which reduces to the linear least-squares when $f(x) = Ax - y$. The problem is solved by making an initial guess for x; that is x_1, and repeating the process for $x_k, k = 1, 2, \ldots$. f is linearizing around x_k using the Taylor series as

$$\hat{f}(x; x_k) = f(x_k) + Df(x_k)(x - x_k). \tag{3.31}$$

D represents derivation. Now, the least-squares problem is solved in iteration via minimizing

$$|f(x_k) + Df(x)(x - x_k)|^2. \tag{3.32}$$

If $Df(x_k)$ has linearly independent columns, the solution for x_{k+1} is computed as

$$x_{k+1} = x_k - (Df(x_k)^T Df(x_k))^{-1} Df(x_k)^T f(x_k). \tag{3.33}$$

The iteration is continued until it converges; that is, $\Delta x_k = x_{k+1} - x_k$ approaches zero,

$$\begin{aligned}\Delta x_k &= -(Df(x_k)^T Df(x_k))^{-1} Df(x_k)^T f(x_k) \\ &= -\frac{1}{2}(Df(x_k)^T Df(x_k))^{-1} \nabla g(x_k),\end{aligned} \tag{3.34}$$

where

$$\nabla g(z) = \begin{bmatrix} \frac{\partial g}{\partial x_{i1}}(z) \\ \vdots \\ \frac{\partial g}{\partial x_{in}}(z) \end{bmatrix} \tag{3.35}$$

An example MATLAB code is provided with this book for testing the Gauss-Newton algorithm for simulated device positioning using only angle measurements, only range measurements, or both range and angle measurements for a set of base stations (or anchor nodes). By default, the accuracies of range and angle measurements are based on CRBs according to adjustable simulation parameters. However, the code can be easily modified to consider also other accuracies or even empirical measurements.

3.3.2 Trilaterion Using Least-Squares Estimation

A basic example of using LS for position computation is trilateration. Trilateration is used for calculating the user's three-dimensional position x from three range measurements, which is the distance between the known positions of three transmitters x_{ti} and the user's receiver as discussed in Section 3.1.1. Trilateration is used in GNSS positioning; however, there the requirement for computing the three-dimensional position is to have four range measurements. Four measurements are required for GNSS as the knowledge that the receiver is close to the Earth may be used as a constraint decreasing the need for measurements to three, but the receiver clock error must be solved simultaneously, increasing the required range measurements to four. Trilateration is also the basis for any positioning system using range measurements. Figure 3.1 showed an example of trilateration with three transmitters emitting ranging signals. Each transmitter provides a range measurement ($y = d_i, i \in 1..n$) to the receiver defining its position at any point on the circle around the transmitter with radius (d_i). When three such circles have been defined based on the three range measurements, the actual user receiver position may be specified to their intersection point. Because range measurements are always noisy, they are called pseudorange measurements. Even if noise was not as large as presented in Section 3.1.1 and all ranging circles would intersect, estimation is required to get an accurate position solution. And again, the more measurements, the better solution is in most cases obtained. The three range measurements are presented as in Section 3.1.1,

$$\sqrt{(x_u - x_{ti})^2 + (y_u - y_{ti})^2 + (z_u - z_{ti})^2} = d_i, \qquad (3.36)$$

where $x = (x_u, y_u, z_u)$ is the user position and $x_t = (x_{ti}, y_{ti}, z_{ti})$ are the known positions of transmitters i.

This is a nonlinear least-squares problem where

$$f(x) = |x - x_{ti}| - d_i. \qquad (3.37)$$

After linearization of and organization of the equations into matrix form, the system is in form Ax = y, where

$$A = 2\begin{bmatrix} (x_{t1} - x_{tn}) & (y_{t1} - y_{tn}) & (z_{t1} - z_{tn}) \\ (x_{t2} - x_{tn}) & (y_{t2} - y_{tn}) & (z_{t2} - z_{tn}) \\ \vdots & & \\ (x_{t(n-1)} - x_{tn}) & (y_{t(n-1)} - y_{tn}) & (z_{t(n-1)} - z_{tn}) \end{bmatrix} \qquad (3.38)$$

$$y = \begin{bmatrix} x_{t1}^2 - x_{tn}^2 + y_{t1}^2 - y_{tn}^2 + z_{t1}^2 - z_{tn}^2 + d_n^2 - d_1^2 \\ x_{t2}^2 - x_{tn}^2 + y_{t2}^2 - y_{tn}^2 + z_{t2}^2 - z_{tn}^2 + d_n^2 - d_2^2 \\ \vdots \\ x_{t3}^2 - x_{tn}^2 + y_{t3}^2 - y_{tn}^2) + z_{t3}^2 - z_{tn}^2 + d_n^2 - d_3^2 \end{bmatrix} \qquad (3.39)$$

and $x = (x_u, y_u, z_u)^T$, for which the solution is obtained using the least-squares method $x = (A^T A)^{-1} A^T y$.

3.4 Fingerprinting

Fingerprinting is a process of comparing perceived measurements with a database of previously collected ones and thereby determining a position. Fingerprinting is divided into two phases: offline and online. In most cases the measurements are indications of signal power level received by an antenna RSSI. RSSI forms the location signature of the antenna position. At the offline phase, also called training or calibration, signal fingerprints for each selected location are created. The locations, called reference points (RPs), may be selected at the room level or more densely, depending on the desired positioning precision. At each reference point (x_n, y_n) the received signal strengths of all signals at the coverage range are measured for the selected time and stored in the database. The online phase is when using the system for navigation, user position (x_u, y_u) is computed by measuring the received signal strength environment and matching that to the stored fingerprints. Figure 3.14 shows the basis of forming a model in the offline phase and using it for positioning in the online phase.

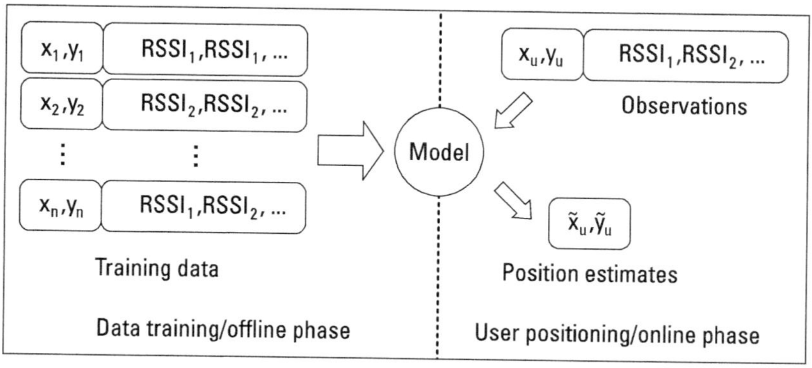

Figure 3.14 A fingerprinting database is created in an offline phase using training data, and then the resulting model is used for computing the user position in an online phase.

Fingerprinting is a popular method for indoor navigation, as the previously discussed signal propagation time-based methods suffer from shadowing and multipath propagation effects that degrade accuracy. Another advantage of using fingerprinting is that it exploits already existing network infrastructures, in most cases WLAN (IEEE 802.11). Recently, other signals characterizing the environment, such as the magnetic field [12], have also started to be used in indoor navigation. Generally, positioning accuracy of a few meters can be achieved, but solutions providing even meter-level accuracy have been presented.

Indoors, many spatial and temporal factors affect the received signal strength. Path loss model defines how the signal power dissipates in an ideal situation where nothing blocks the signal, as a function of the distance (d) between the AP transmitting the signal (G) and the receiver (R)

$$P_R = P_T G_R G_T (\frac{\lambda}{4\pi d})^2. \tag{3.40}$$

where P_R and P_T are the received and transmitter powers, respectively, G_R and G_T are the gains of the receiving and transmitting antennas, and λ is the signal's wavelength. Figure 3.15 shows an example of the RSSI in a more realistic indoor environment, where the signal strengths do not dissipate with respect to the transmitter distance but get degraded due to the obstacles.

3.4.1 Creating the Database

Locations of the reference points must be measured and saved into the database. The simplest approach is to manually select the reference point locations on a floor plan. If the floor plan is connected to an established coordinate system, the actual location coordinates may be estimated at the navigation stage. One solution for creating the fingerprinting database is to consider the navigation

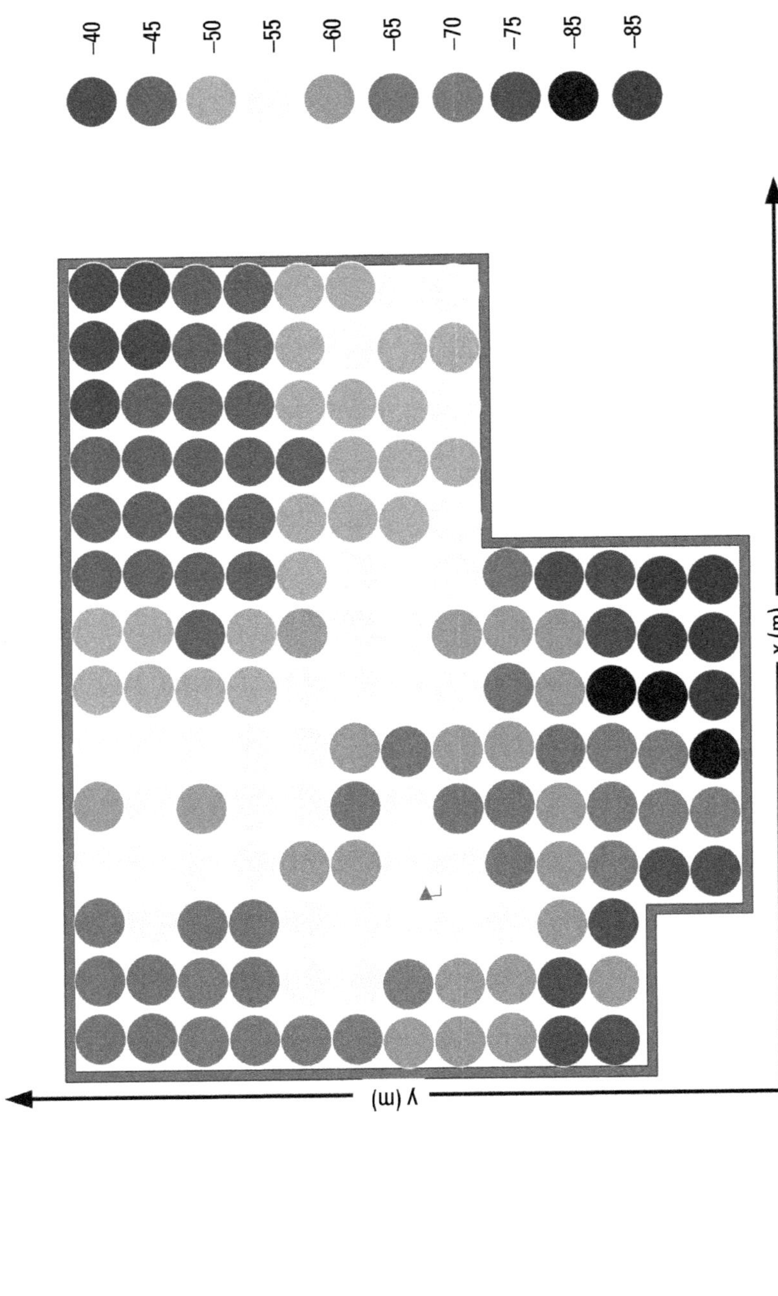

Figure 3.15 Signals in an indoor environment, where the signal strengths get degraded due to obstacles. The transmitter is the small triangle.

environment being composed of cells with reference points as their centers from which the training RSSI values for all APs in the coverage range are collected [13]. The fingerprint measurement R_i for the reference point $i, i = (1, \ldots, p)$ may be presented as conditional probability distributions $P(AP_m RSSI_r | R_i)$ for each AP (AP_m) on the coverage range $(1, \ldots, k)$ [14].

$$\begin{bmatrix} P(AP_1 RSSI_1 | R_i) & \ldots & P(AP_k RSSI_1 | R_i) \\ P(AP_1 RSSI_2 | R_i) & \ldots & P(AP_k RSSI_2 | R_i) \\ \vdots & \vdots & \vdots \\ P(AP_1 RSSI_v | R_i) & \ldots & P(AP_k RSSI_v | R_i) \end{bmatrix} \tag{3.41}$$

The probability of measuring the RSSI value $RSSI_r, r = (1, \ldots, v)$ transmitted by the AP_m at a reference point i when measuring the signal environment as R_i is

$$P(AP_m RSSI_r | R_i) = \frac{C_{RSSI_r}}{N_i}, \tag{3.42}$$

where C_{RSSI_r} is the occurrence of $RSSI_r$ in the database training set and N_i is the total number of training samples collected at the reference point i. The fingerprinting database D is then compiled from the fingerprint measurements R_i from each reference point as $D = [R_1, R_2, \ldots, R_p]$.

3.4.2 RSSI-Based Positioning

At the navigation phase, the user perceives the signal environment and compares the measured RSSI values with the ones in the database and uses an algorithm for evaluating the accuracy of the match.

Euclidean distance (d_s) is a simple method for computing the difference between the RSSI values measured at the estimated user position (x_u, y_u) and the database indicated logarithmic signal strength at RP [15]

$$d_s(x_u, y_u) = \sqrt{\frac{1}{m} \sum_{i=1}^{m} [RSSI_i - RSSI_{di}(x_u, y_u)]^2}, \tag{3.43}$$

where $RSSI_i$ is measured in decibels and $RSSI_{di}$ is the signal strength value in the database for the estimated user position. m is the number of signals in the database at the estimated position coverage range.

A more sophisticated algorithm for evaluating the accuracy of the match is K-weighted-nearest neighbors (KWNN). The KWNN algorithm uses (3.43) to select the K-nearest RPs with weighted average for position calculation. Because the distances from the selected reference points to the estimated user position are different, so are the weights assigned to the reference points. The weights of the selected reference points are calculated as

$$w_{is}(x_u, y_u) = \frac{1}{d_s(x_u, y_u)} \sum_{i=1}^{k} \frac{1}{d_{is}(x_u, y_u)}, \quad (3.44)$$

and finally the improved estimation for the user position $(\tilde{x_u}, \tilde{y_u})$ as

$$\tilde{x_u}, \tilde{y_u} = \sum_{i=1}^{k} w_{is}(x_u, y_u), \quad (3.45)$$

One of the strengths of fingerprinting is that it does not require the installation of any new infrastructure or the modification of existing ones compared to, for example, triangulation from TOA requiring precise synchronization among all transmitters and receivers. However, the major challenge related to fingerprinting is the effort required to build and maintain high-quality radio maps. Nonstandardized localization hardware and thereby receiver-related differences affect the observed signal power [16]. Also, RSSI measurements' dependency on the device and measurement position distort the solution. The requirement to link the reference point to a real physical location necessitates the availability of a map, floor plan, or another offline position measuring method, which are not always available. Dynamic changes in the navigation environment, such as layout changes and moving people, affect the signal environment. Although fingerprinting is less sensitive to multipath and fading it still affects the signal power estimation accuracy. Finally, the limited precision and/or accuracy of the positioning estimation algorithms degrade the navigation solution. Currently, device and user dependent challenges as well as the need for decreasing the workload in forming the database have been addressed via the use of crowdsourcing. Crowdsourcing means that multiple people attend the fingerprinting process by recording the RSSI measurements and tagging those with the correct location.

3.5 Dead Reckoning

Dead reckoning (DR) is a relative positioning method. If the initial position is known, the current position may be determined by propagating the measured distance and direction traveled. Figure 3.16 shows a two-dimensional dead reckoning process. The known initial position (p_0) at time epoch t_0 has been propagated with four distance (d) and direction (ω) measurement updates, eventually resulting in position (p_4) at time epoch t_4. The measurements for dead reckoning may be provided by an inertial navigation system and turned into distance or speed via strapdown computation as discussed above. Relative position p_{t_k} for each time epoch $t_k, k \in]0, \ldots, N]$, where N is the final epoch

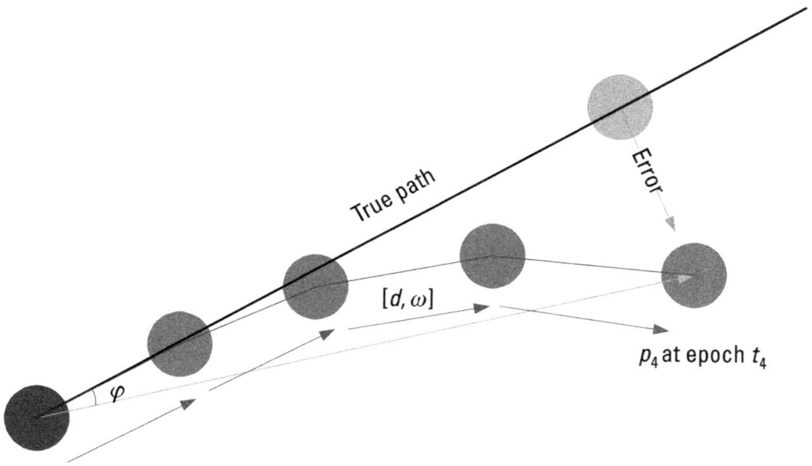

Figure 3.16 DR-based navigation solution, where the initial position is propagated using sensor measurements. Due to measurement errors, the position solution begins to drift and will result in position error and heading drift shown with angle ϕ.

of navigation, is simply obtained using $p_{t_k} = p_{t_{k-1}} + d_{t_k}$. Direction is used for transforming the relative position solution p_{t_k} into absolute position p_k in the correct coordinate system.

As the measurements used in DR are usually measured in the body frame; coordinate transforms are usually required for position propagation. The two-dimensional solution may be extended into three-dimensional by considering the full orientation solution instead of only the heading. The benefits of DR compared to absolute position determination are continuous operation, often a high update rate of the measuring system, low short-term noise, and the provision of, for example, attitude in addition to position (and sometimes velocity).

Figure 3.16 also shows the challenge in computing a DR-based navigation solution. The position error growth caused by sensor artifacts is unbounded. As the measurements of distance and direction are relative, even a small error causes *drift* in the solution. Traditionally, navigation drift was related to the error in the heading and presented with drift angle (ϕ), but at present drift is usually related to the resulting position error. Drift (i.e., position error) grows with time because the errors in successive distance and direction measurements accumulate. The accuracy of the solution is dependent on the measurement sensor's quality, but even with the best-quality systems drift is inevitable. Because the uncertainty of dead reckoning increases over time and distance, for longer-term navigation the solution must be corrected and sensor errors calibrated occasionally with a position fix from an absolute positioning system. Indoors, radio positioning is often used as the absolute system for correction; however, simultaneous localization and mapping obtained using cameras and computer vision embodies a method for drift correction, called loop-closure, which is discussed in Chapter 4.

Dead reckoning is straightforward to be embedded into statistical filtering, for example Kalman filtering. This applies to the motion of a pedestrian in most cases: the stepwise movement usually takes the direction where the person is facing. The occurrence of a step can be detected from inertial measurements; accelerometers are plotted in Figure 3.17, but gyroscope signals are also highly correlated with the steps. In the simplest form, assuming the average step length of the user is known or can be modeled adequately, then steps can be detected from an accelerometer mounted on the torso of the user (Figure 3.17(a)) while gyroscopes are used to estimate the direction of travel.

However, the step length of a person is challenging to predict and often needs user-specific calibration parameters. A more general solution can be implemented by mounting the IMU on the foot of the user: during the foot stance phase, the IMU is almost stationary as can be seen from the accelerometer in Figure 3.17(b). In the following section we show how this property can be utilized.

3.5.1 Pedestrian Dead Reckoning

Drift in an INS-based DR system is substantial when the size and cost requirements necessitate the use of low-quality MEMS sensors, as in pedestrian navigation. The third column of Table 3.1 presents the magnitude of horizontal position error obtained after 10 seconds of navigation when using dead reckoning and inertial sensors of different grades (navigation, tactical, and consumer) characterized by gyro bias shown in the second column [17]. The solution obtained after 10 seconds of strapdown MEMS inertial sensor navigation using dead reckoning has 100m of position error, which is completely unacceptable for practically all applications. However, pedestrian dynamics provide information that can be used as a constraint for reducing the error growth [18]. During each step, the foot alternates between a stationary *stance* phase and a moving *stride phase*, each lasting about 0.5 seconds [19]. Traditionally, pedestrian dead reckoning (PDR) has been done by calculating the pedestrian's step count by detecting the phases [15], but such methods require the knowledge of user's step length and are not very practical. The stationary phase, when the system has a constant position and attitude, can be used to bound the error growth when the sensor is placed on the pedestrian's foot. This method is called zero-velocity update (ZUPT). The fourth column of Table 3.1 shows significant decrease in the position error obtained when ZUPT is used in navigation; the process is referred to as foot-mounted pedestrian dead reckoning.

Two types of ZUPTs exist; hard and soft updates [18]. Soft ZUPTs are used when the foot motion measurements are incorporated into a navigation system based on statistical filtering and the PVT solution is computed over time.

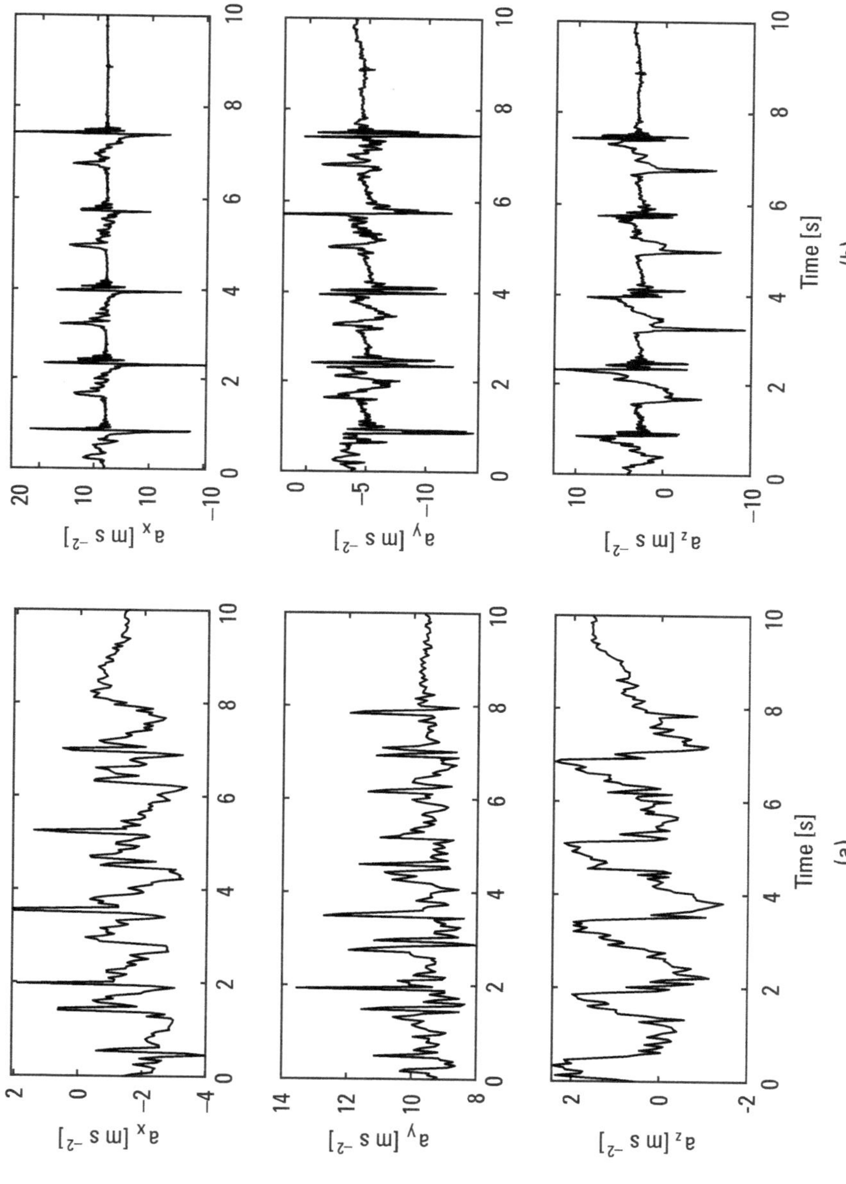

Figure 3.17 Accelerometer measurements from a pedestrian user: (a) IMU mounted on the waist, and (b) IMU mounted on the foot.

Table 3.1
Inertial Navigation Errors

		Error in Horizontal Position (m)	
IMU Grade	**Bias (deg/h)**	**No ZUPT**	**ZUPT**
Navigation	0.01	0.05	0.0001
Tactical	1	0.5	0.001
Consumer	100	100	0.5

Such a setup provides an estimate of how errors develop over time and provides an estimate of the accumulated errors since the last zero-velocity update [19]. Therefore, the previously discussed strapdown mechanization can essentially be interpreted as a boundary value instead of an initial value problem and the step displacement may be computed [20] using

$$\begin{aligned} \dot{R} &= R\left[\omega^B\right]_\times \\ \dot{v}^L &= Rf + g^L \\ \dot{p}^L &= v^L \end{aligned} \tag{3.46}$$

where the matrix R is the rotation from the body frame B to the local navigation coordinate frame L; $[]_\times$ denotes the 3×3 skew-symmetric cross product matrix; v^L and p^L are the velocity and position in the navigation frame, respectively; and g^L denotes the local gravitational acceleration vector. Note that here certain factors from the conventional strapdown mechanization that are not significant in pedestrian navigation, such as the transport rate and the Coriolis force, are neglected. The estimate is then fed back to the filter to correct the navigation solution. In foot-mounted PDR it is straightforward to apply a ZUPT to the error-state filter whenever the IMU is detected to be at rest. The measurements justify the use of the term PDR in the context of the strapdown navigation although traditionally PDR and strapdown mechanization have been used exclusively. However, note that the displacement measurement in foot-mounted PDR is a three-dimensional vector instead of a scalar quantity. Hard ZUPT provides corrected measurements of the foot motion during an individual gait [21]. At the update, position, velocity, and yaw are reset to zero, and roll and pitch are initialized using measured gravity acceleration.

The most critical part of ZUPT-based PDR is to detect the stationary periods correctly. The objective of the zero-velocity detection is to decide whether the foot is moving or stationary during a time epoch consisting of N (depending on the sensors sample rate and the length of the stationary period) observations from the inertial sensors between the time instants n and $n + N - 1$. The different stance detectors found in the literature are variation or magnitude of acceleration and angular rate energy [22], the former using only accelerometer and the latter only gyroscope data. A generalized likelihood ratio test (GLRT), called stance hypothesis optimal detection (SHOE) [18], has shown good performance on stance detection. GLRT is based on evaluating whether the logarithm of likelihood ratio ($T(z_n)$) is below a defined threshold γ as

$$T(z_n) = \frac{1}{N} \sum_{k \in \Omega_n} \left(\frac{1}{\sigma_a^2} \| a_k - g \frac{a_n}{\| a_n \|} \|^2 + \frac{1}{\sigma_w^2} \| w_k \|^2 \right) \prec \gamma, \quad (3.47)$$

where a_k and a_n are the measurements of specific force and angular rate at time epoch k, σ_a and σ_w denote the accelerometer and gyroscope noise variance, respectively, Ω_n is the set of indexes over which the averaging is done, and g is the magnitude of the local gravity vector, $a_n = \frac{1}{N} \sum_{k \in \Omega_n} a_k$. SHOE and (3.47) can be interpreted as follows. The first squared term inside the parenthesis forms the mean square error of fitting a vector of magnitude g with the direction of the average specific force vector to the accelerometer data. The second term is the energy in the gyroscope signal, which is combined with the former term after both terms are weighted by the quality of the measurements. If the result is smaller than the selected threshold γ a stance phase is detected. Figure 3.18 shows the values of acceleration, angular velocity, and SHOE detector during step phases.

Selection of the threshold is also critical and a simple threshold fails when the motions are unusual, such as running and climbing. Adaptive thresholding has provided improvement for the detection. An adaptive threshold considering both velocity and acceleration [23] has shown good results when running and crawling; stages of the pedestrian's gait may be detected using a hidden Markov model [24]. Currently, machine learning has emerged in PDR development for modeling the motion modes of the pedestrian and utilizing the information in thresholding of stance phase, for example running or walking [25] and also climbing [26].

3.6 Time Series Estimation

In most positioning applications, measurements are obtained at regular intervals. Typically, some prior knowledge on the time series of position estimates is available: for instance, the velocity of a pedestrian in a shopping mall rarely exceeds several meters per second. The purpose of statistical filtering is to combine such prior knowledge on the evolution of the position (or other unknowns) with the actual measurements. In the following sections, we introduce the principles of Bayesian filtering. Then, some popular filtering algorithms are presented. In the scope of this book, we limit the discussion to discrete-time models and algorithms.

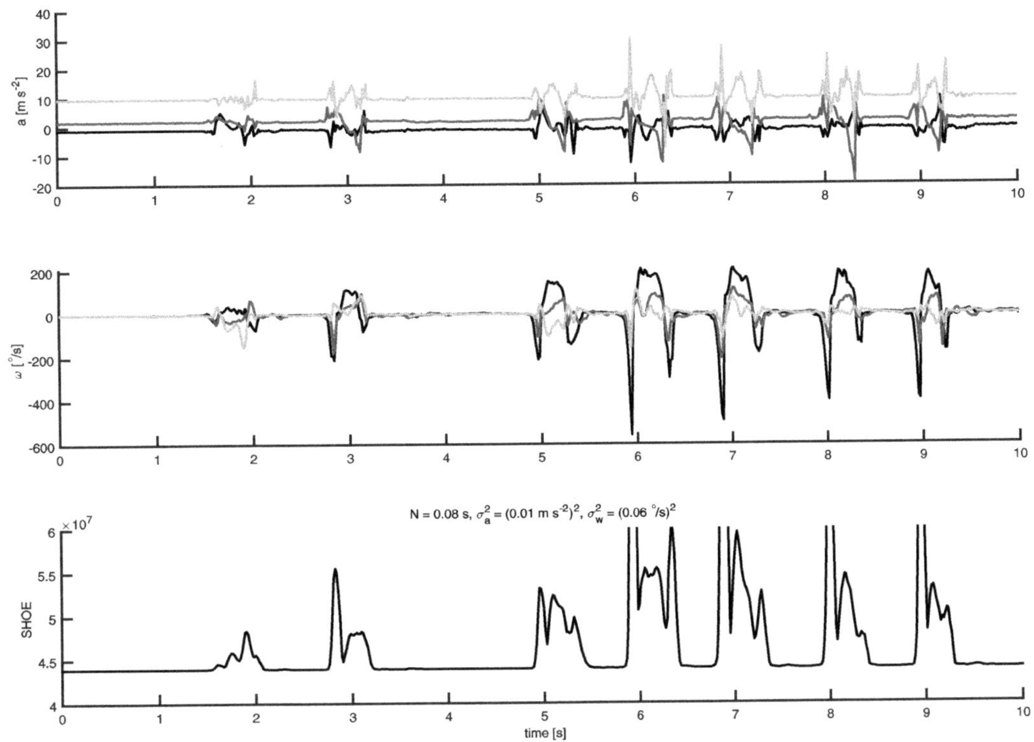

Figure 3.18 Step phases detected using acceleration (specific force), angular velocity, and a SHOE detector. During the stance phase, angular velocity and acceleration on the x- and y-axes are close to zero, acceleration on the vertical z-axis is close to the local gravity magnitude, and SHOE value is below the predefined threshold.

3.6.1 Bayesian Filtering

In statistical filtering, the unknowns are called *state variables*, conventionally combined into a state vector (often called simply the state). The goal of statistical filtering is to estimate the *distribution* of the state variables; often the state is represented in terms of summary statistics such as the mean value and covariance matrix to quantify the uncertainty. Thus, as opposed to static estimation (e.g., Section 3.3), in the filtering context the vector of unknowns is associated with a time index t, resulting in the state vector x_t. The state vector is associated with a (discrete-time) state transition function f that allows predicting the state distribution at time $t + 1$ given the current estimate:

$$x_{t+1} = f(x_t) + \eta_t \tag{3.48}$$

where η_t denotes the state transition uncertainty, often called *process noise*. It is common to assume the state transition process to be *Markovian* (i.e., that the transition to x_{t+1} only depends on x_t and not any earlier values). It is also possible to include a control input to the state model to express deterministic state variations that stem from external actions (e.g., a robot moves because the operator commanded it to), but this is omitted from (3.48) for simplicity.

In general, the state transition function f and the state variables in x are chosen such that the process noise η becomes a zero-mean random variable. In Bayesian filtering, the transition model allows the computation of the conditional probability $P(x_t|x_{t-1})$; to initialize the filter, an initial state distribution $P(x_0)$ is needed. The initial state distribution can be obtained, for example, using a static algorithm or other prior information.

The state transition model is based on a priori information. Actual observation data is incorporated by means of measurement updates of the form

$$y_t = h(x_t) + n_t \tag{3.49}$$

where h is the measurement function and n_t represents the measurement uncertainty that is again assumed to have zero mean value. Furthermore, it is assumed that the measurement noise n_t is independent of the value of the state x_t; in contrast, the measurements in y_t can be mutually correlated as long as this is accounted for in the distribution of n_t. The measurement updates enable the computation of the conditional probability $P\left(x_t|y_t\right)$. Combined with the state transition model, the estimation can take the entire measurement history into account, resulting in the filtered state distribution $P\left(x_t|y_{1,\ldots,t}, x_0\right)$. Filtering algorithms generally conduct the time series estimation recursively (i.e., by applying the new information on the previous estimate without storing the history of measurements explicitly).

It is noteworthy that the measurement function h does not need to depend on all components of the state vector x: the state vector distribution estimated by the filter can include cross-correlation between individual state variables (originating from the state transition function f), which allows the estimation of unknowns that are not directly related to the measured quantities. A state-space model where all unknowns can be estimated is said to be *observable*; the observability of a system can be analyzed based on the transition and measurement functions [27].

3.6.2 Kalman Filtering

The Kalman filter (KF) [28] is the minimum variance solution for the Bayesian filtering problem in the special case where the state transition and measurement models are linear with Gaussian uncertainty; that is,

$$\begin{aligned} x_{t+1} &= \mathrm{F}x_t + \eta_t, \quad \eta_t \sim \mathcal{N}(0, \mathrm{Q}_t) \\ y_t &= \mathrm{H}x_t + n_t, \quad n_t \sim \mathcal{N}(0, \mathrm{R}_t) \end{aligned} \tag{3.50}$$

where $\mathcal{N}(\boldsymbol{\mu}, \Sigma)$ denotes the normal distribution with mean $\boldsymbol{\mu}$ and covariance Σ. The matrices F and H need not be constant in time, but time indices are omitted here for simplicity. A normal distribution is completely characterized by the mean value and the covariance matrix, which allows the KF to work efficiently by only maintaining estimates of these quantities. In this book we will not go through the complete derivation of the Kalman filter equations; the interested reader is referred to [29].

With the KF assumptions, the state transition is evaluated as

$$P(x_{t+1}|x_t) = \mathcal{N}\left(\mathrm{F}\boldsymbol{\mu}_t, \mathrm{F}\Sigma_t\mathrm{F}^T + \mathrm{Q}_t\right) \tag{3.51}$$

where we use $\boldsymbol{\mu}$ to denote the mean value of the state distribution. It is common in literature to use x to represent both the (unknown) state vector and its mean value, but using a separate symbol for the mean emphasizes the fact that we are estimating a complete distribution, not just a single value.

The measurements at time t are used to update the state distribution as follows:

$$\begin{aligned} P\left(x_t|y_{1,\ldots,t}, x_0\right) &= \mathcal{N}(\boldsymbol{\mu}_t + \mathrm{K}_t z, (\mathrm{I} - \mathrm{K}_t\mathrm{H})\Sigma_t) \\ \mathrm{K}_t &= \Sigma_t\mathrm{H}^T\left(\mathrm{H}\Sigma_t\mathrm{H}^T + \mathrm{R}_t\right)^{-1} \\ z &= y_t - \mathrm{H}x_t \end{aligned} \tag{3.52}$$

where I is the identity matrix, and the matrix K is known as the *Kalman gain* (i.e., the relative weight of the measurement information with respect to the prior estimate). The vector z represents the difference between the actual measurement and the value predicted based on the state estimate, commonly called the *innovation*. A KF can be severely disturbed by gross measurement errors; to improve the robustness, outliers can be detected and rejected by, for example, monitoring the innovation values [30].

As an example, Figure 3.19 compares the static least-squares position estimates and the Kalman filtered time series where the positioning is based on ranging to three base stations. The bivariate state vector consists of the x- and

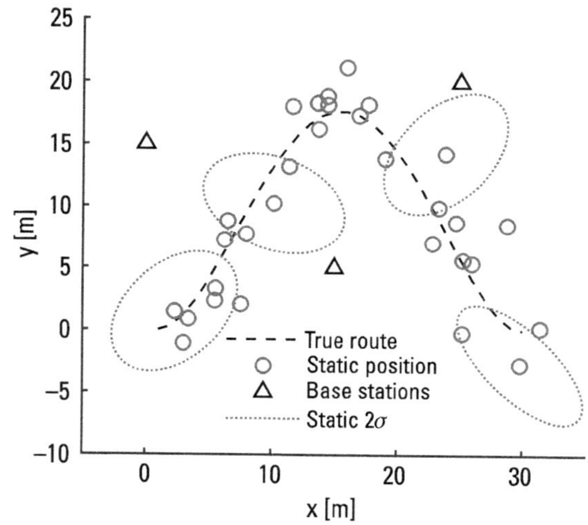

(a)

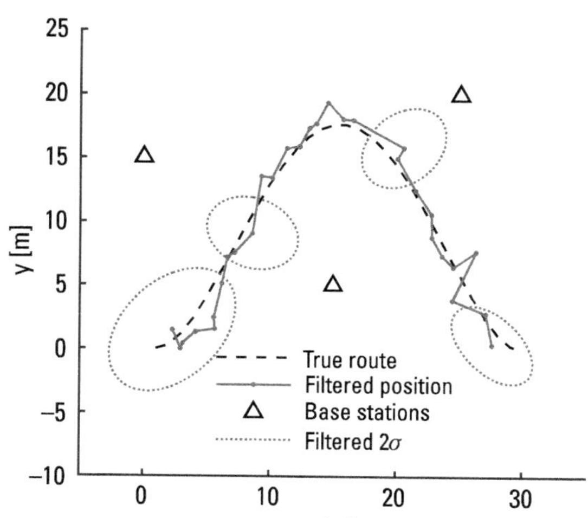

(b)

Figure 3.19 Comparison of static positioning (a) and a Kalman filtered time series estimate (b) in the scenario of 2-D ranging with three base stations. The direction of motion is from left to right.

y-coordinates, and the filter model is as follows:

$$
\begin{aligned}
F &= I \\
Q &= (1.5\ \mathrm{m})^2\, I. \\
H &= I
\end{aligned}
$$

This model assumes that the user is moving at a speed in the order of $\sqrt{2} \times 1.5$ m/s with randomly fluctuating heading. The measurement y_t is the static position estimate determined at time t using the Gauss–Newton algorithm (see Section 3.3) and the measurement covariance R_t is obtained by (3.30) assuming 2m standard deviation for the ranging errors. Figure 3.19 also shows the 2-sigma confidence ellipses at 10-second intervals; it can be seen that the uncertainty of the filtered solution decreases over time relative to the static solution. Note that the mutual geometry of the user and the base stations affects the instantaneous positioning error covariance, which is why the ellipses change their shape over time.

In general, the KF is computationally efficient and widely applied in various navigation solutions. However, the assumption of linear state transition and measurement models is quite strict, which is why a nonlinear variant known as the EKF is commonly used instead. The nonlinear extension is based on linearizing the state transition and measurement functions f and h, respectively, which implies the following modifications to the Kalman filter equations (3.51) and (3.52):

$$
\begin{aligned}
F &= \left.\frac{\partial f}{\partial x}\right|_{x=\mu_t} \\
P\left(x_{t+1} \mid x_t\right) &= \mathcal{N}\left(f(\mu_t),\ F\Sigma_t F^T + Q_t\right). \\
H &= \left.\frac{\partial h}{\partial x}\right|_{x=\mu_t} \\
z &= y_t - h\left(\mu_t\right)
\end{aligned}
\tag{3.53}
$$

The EKF works remarkably well in cases where the state transition and measurement models are close to linear and the uncertainties can be modeled as Gaussian. In a generic navigation context, GNSS pseudoranges are a classical example of measurements well suited to the EKF framework; since the satellites are 20,000 km away, the pseudoranges behave locally like linear functions although the measurement model is quadratic. Unfortunately, situations where the EKF assumptions simply do not apply are rather common in indoor navigation. In the following section we introduce a filtering method that is free of these limitations.

3.6.3 Particle Filtering

Kalman-type filters assume the underlying distributions to be Gaussian, which allows the estimates to be maintained in terms of a mean vector and a covariance matrix. If the system is too complex to be modeled with normal distributions, one alternative is to use Monte Carlo samples to represent the state distribution estimate. A particle filter (PF) uses a set of N random state vector samples $\boldsymbol{x}^{(i)}$, $i = 1, \ldots, N$, with associated weights $w^{(i)} \in [0, 1]$ to capture the characteristics of the state distribution. The state vector samples $\boldsymbol{x}^{(i)}$ are referred to as *particles* to emphasize that they represent hypotheses rather than measured samples. This format can represent arbitrary probability distributions, as illustrated in Figure 3.20; note that in Figure 3.20(b), the two range measurements are not sufficient alone to determine a unique position solution, as discussed in Section 3.1.1.

The particles are generated by drawing N random samples from a *proposal distribution* $p^*\left(\boldsymbol{x}_t|\boldsymbol{x}_{0,\ldots,t-1}, \boldsymbol{y}_{1,\ldots,t}\right)$. When implementing a particle filter, the proposal distribution is selected to approximate the true state distribution $P\left(\boldsymbol{x}_t|\boldsymbol{y}_{1,\ldots,t}, \boldsymbol{x}_0\right)$, yet it should be easy to draw samples from the proposal distribution [29]. The fact that the proposal distribution is just an approximation is compensated for by adjusting the particle weights according to the measurement. This procedure is conducted as follows:

$$\begin{aligned} &\boldsymbol{x}_t^{(i)} \sim p^*\left(\boldsymbol{x}_t|\boldsymbol{x}_{0,\ldots,t-1}^{(i)}, \boldsymbol{y}_{1,\ldots,t}\right) \quad \text{for } i = 1, \ldots, N \\ &w_t^{(i)} \propto w_{t-1}^{(i)} \frac{P\left(\boldsymbol{y}_t|\boldsymbol{x}_t^{(i)}\right) P\left(\boldsymbol{x}_t^{(i)}|\boldsymbol{x}_{t-1}^{(i)}\right)}{p^*\left(\boldsymbol{x}_t^{(i)}|\boldsymbol{x}_{0,\ldots,t-1}^{(i)}, \boldsymbol{y}_{1,\ldots,t}\right)} \text{ with } \sum_{i=1}^{N} w_t^{(i)} = 1 \end{aligned} \tag{3.54}$$

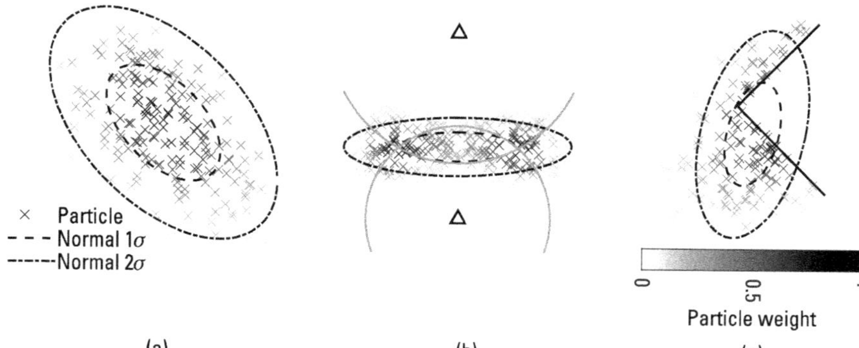

Figure 3.20 Examples of probability distributions in the weighted particle representation and the normal distribution fitted to the samples: (a) pure normal distribution, (b) bimodal distribution arising from ranging to two base stations, and (c) distribution with a map constraint.

The choice of the proposal distribution $p^*\left(x_t|x_{0,\ldots,t-1}, y_{1,\ldots,t}\right)$ has an effect on the filter performance: if it is not a close approximation of the true state distribution, a higher number of particles N may be needed, resulting in increased computational burden. For instance, a common PF variant called the *bootstrap filter* is obtained by setting the proposal distribution to be the same as the transitional model (i.e., $p^*\left(x_t|x_{0,\ldots,t-1}, y_{1,\ldots,t}\right) = P(x_t|x_{t-1})$). The bootstrap filter is very simple to implement, but it is obvious that this choice of proposal distribution is far from optional as it does not take the most recent measurement into account at all. For instance, the *auxiliary particle filter* [31] generates a more accurate proposal distribution by adding an auxiliary variable to the PF algorithm.

Repeated iteration of (3.54) over time will eventually lead to a situation where most of the particles have a negligible weight, with just a few particles contributing to the estimation with a significant weight. Such a situation is obviously undesirable as the set of particles loses diversity, and maintaining particles with negligible weight is a waste of computational resources. This situation can be avoided by resampling a new, uniformly weighted set of particles representing the same distribution. A common method is *stratified resampling* [32] where the new particles are obtained by drawing from the discrete distribution constituted by the old particles and their weights: in other words, the probability of the new particle $x^{(i)}$ being equal to the old $x^{(j)}$ is $w^{(j)}$. Once the resampling is done, the weights are reset to $w^{(i)} = 1/N \ \forall i$. To optimize computational resources, the particles need not be resampled at every measurement epoch but, for instance, only if the variance of the set of weights exceeds a certain threshold [33].

3.6.4 Factor Graph Optimization

In recent years, factor graph optimization (FGO) has gained interest in the navigation research community. While Bayesian filtering considers the measurement and state transition history to be captured in the estimated state distribution, FGO uses an optimization algorithm to determine the entire time series of state values, with the cost function including all the measurements and other constraints [34].

The name of the technique refers to factorizing the (global) cost function into the form of a graph. The graph includes two types of nodes: *factors* (or local functions), which correspond to constraints, and *variables*, which are the unknowns to be optimized. Edges only exist between nodes of different types: an edge indicates a dependency between the factor and the variable.[1] A very simple factor graph, consisting of an initial state x_0 and two subsequent epochs x_1, x_2 with corresponding measurements y_1 and y_2, is sketched in Figure 3.21. For

1. Different graphical representations to factor graphs exist. For instance, in some texts, the variables are denoted with edges (which may be connected to one or two nodes) [35].

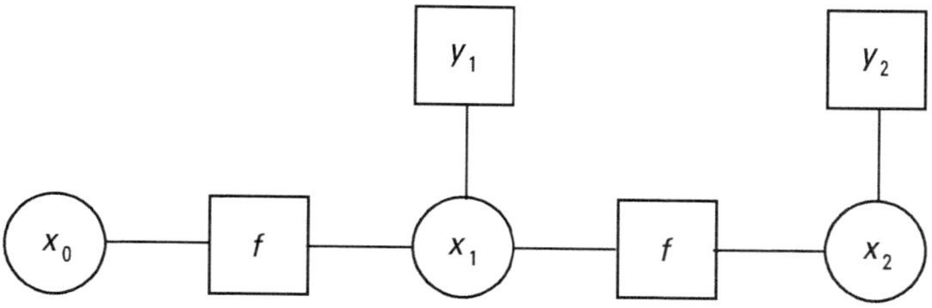

Figure 3.21 Factor graph representation of the state-space navigation model.

navigation problems it is typical that the factor graph has a chain-like structure, but this is not always the case; for instance, in SLAM, a single landmark may be observed from several locations at different times, connecting the corresponding landmark position variable to several user position variables via observation factors [34].

In FGO, the global cost function is minimized by minimizing the product of the factors. Depending on the functions involved in the factors, this may be possible through, for example, linear least squares estimation or an algorithm such as Levenberg–Marquardt optimization. However, in practical applications, it is desirable to compute the solution incrementally so that the estimate is updated whenever new information is obtained without computing a new batch solution at each epoch. This can be done, for example, using the iSAM2 algorithm [36].

3.7 The Future of Navigation Algorithms: Machine Learning

Machine learning (ML) is the process of using mathematical models of data to infer and predict different phenomena without hardcoding the rules. This section presents some of the interesting solutions to build an understanding of the capabilities ML provides for navigation. First the three classes of ML are presented, which are unsupervised, supervised, and reinforcement learning. Then, ML-based indoor navigation applications are discussed.

3.7.1 Unsupervised, Supervised, and Reinforcement Learning

As stated above, machine learning methods are divided into three categories: supervised, unsupervised, and reinforcement learning. The goal of supervised learning is to develop a model that learns a function used to predict the value of a variable y by using input data x. Predictions are of two types: regression, where the y is a continuous function, or classification, where y is a categorical variable. In both cases, the known true values of y, which is the ground truth, are called labels and they are used for training the model to predict the value of y also from unseen x. Unsupervised learning does not have the labels of y but is used for developing a model that is able to discover hidden, underlying structure within the data x. One form of unsupervised learning is clustering, where similar data points are clustered together automatically based on their characteristics. However, at present more and more unsupervised ML methods are developed for more sophisticated use cases, such as visual odometry [37]. Figure 3.22 illustrates the difference between supervised and unsupervised learning. On the left is an example of supervised learning, where data points with labels are classified, and on the right is unsupervised clustering of data with hidden underlying structure.

The development of reinforcement learning (RL) methods has been going on for decades, but only recently has their use in various important application fields accelerated. Surprisingly, RL is still not widely used in navigation, although it is based on the Markov decision process (MDP) principles just like particle filtering. RL is a learning process, where an *agent* takes a sequence of *actions* to change its *state* in an environment, and via a *reward* (that could be considered as a penalty too) based on how beneficial the action was for its overall goal, learns a model to reach a favorable solution. Traditionally, RL has been used for playing games [38], but more and more it is also used for improving solutions from traditional ML and different simulations. Methods in all three categories of ML can be based on traditional or deep learning.

3.7.2 Machine Learning for Indoor Navigation

This section discusses some interesting indoor navigation applications of machine learning. The number of applications is increasing rapidly, so the objective for

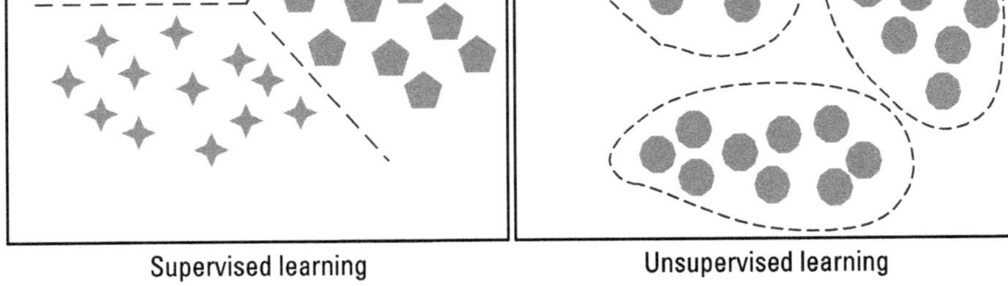

Figure 3.22 Machine learning for classifying data points using supervised learning and data with labels on the left, and for clustering data using unsupervised learning and data with hidden underlying structure on the right.

selecting these applications was to present the strengths and capabilities of some ML algorithms and provide ideas for navigation problems to which they provide solutions.

Although ML research has been active for decades, integration of ML methods into the navigation domain accelerated in the 2010s. The main enabler of integration has been the emergence of deep learning methods discussed already in Section 2.3 which in turn has been accelerated by the increase in publicly available materials. Deep learning is able to process data from complex nonlinear systems and get improved positioning in situations where human perception fails. A representative navigation example is fingerprinting, where the input data is nonlinear and has complex boundaries, making the result inaccurate using conventional computing means. Also, traditional ML algorithms require human guidance in selecting the features to be used in learning, whereas deep learning will observe them from data automatically and therefore is able to use fine details and complex phenomena that are beyond human understanding. An example of indoor navigation using geomagnetic field fingerprinting is presented in [39], where CNN is used to join a pedestrian's observed magnetic fields values with ones in the database and thereby obtain position information.

Deep learning provides relief for synchronization and calibration requirements in measurement fusion, mainly used by fusing visual and inertial measurements with a recurrent neural network (RNN) [40] or with a network fusing both recurrent and convolutional ones (RCNN) [41]. As well, a rapidly emerging into navigation field is a concept called transfer learning, which enables transferring a model trained in a certain navigation environment into a new but similar environment without starting the process from scratch [42]. Machine learning is used at various process phases and for various tasks throughout the navigation process pipeline. It can be used for measurement error detection, improving position computation as an augmentation method or providing the full solution, as for example simultaneous localization and mapping discussed in Chapter 4, understanding the system dynamics and thereby thresholding computation equations, and so forth. Below, we will look at three representative examples of recent ML navigation in more detail. As discussed, the methods are vast and new ones emerge with tremendous speed. Therefore, the selection criteria for the methods discussed was to show cases using both various techniques discussed in the book and different deep learning architectures.

3.7.2.1 UWB-Based Positioning Using Deep Reinforcement Learning

UWB provides an accurate indoor navigation system at a feasible environment. However, moving objects, multipath, sensor noise, changes of environment, and the unmodeled user dynamics degrade the result. A computationally efficient and robust robot navigation solution was developed using UWB positioning and deep RL (DRL) [43]. A simulated environment encapsulating robot dynamics was created and the model trained using distance and angle to goal as observations, obtained with a lidar, odometer, and magnetometer. Thereby, an agent that was able to navigate in the indoor environment without explicitly modeling the UWB sensor setup during training was created and generalizable to the task using only the UWB sensors for navigation. The actions in the model were 2-D vectors containing normalized angular and linear velocities of the robot. The robot was rewarded largely when it reached the goal, modestly if its distance and difference in heading toward the goal was decreasing, and punished if it collided. The experiments showed that in highly complex indoor environments, RL provides tools for automatizing the laborious setup of infrastructure and improving the resulting navigation accuracy.

3.7.2.2 Deep Learning for Sensor Fusion

Traditionally, statistical filtering such as Kalman and particle filters have been used for sensor fusion providing improved navigation accuracy and continuity. Inertial sensors and cameras are often fused due to their different error and degradation sources. However, the challenges in measurement time synchronization and IMU and camera alignment, in addition to the sensor-related deficiencies, add errors to the resulting position solution. Deep learning algorithms—more specifically recurrent neural networks that also consider the time dimension, provide improved fusion accuracy in addition to solving some of the challenges in sensor synchronization and calibration. Optical flow based motion solution was fused with inertial sensor measurements using a long short-term memory (LSTM) recurrent neural network [44] for providing an accurate seamless pedestrian indoor-outdoor navigation solution. Reference trajectory was used to train the model to transform the motion information obtained with the sensors to the composited position solution.

3.7.2.3 Wi-Fi Fingerprinting and Machine Learning

At present, most of the existing Wi-Fi fingerprinting solutions do not scale well for multibuilding (e.g., a shopping mall) environments and consider multifloor positioning as a hierarchical setup, namely map each floor separately and form an integrated solution from those. Deep learning provides tools for less laborious solutions with reduced parameter tuning requirements and adaptability compared to more traditional methods. Scalable indoor localization method based on deep neural networks, a stacked autoencoder (SAE), was presented in [45]. The method provided a feedforward multilabel classifier for detecting the location on a building and floor level. It did not require the traditional hierarchical classification due to using flattened one-dimensional combined building and floor identifiers as labels for training the network.

The bottleneck of supervised learning, especially deep learning, is the requirement for a large amount of labeled data. Therefore, unsupervised learning would be beneficial in a navigation domain as well, but the accuracy of unsupervised methods is still far from corresponding supervised algorithms. At present, unsupervised learning is used mainly in computer vision based navigation and for improving the user position estimate or handling crowdsourced RSSI data in Wi-Fi fingerprinting. The weighted-k-nearest-neighbors algorithm presented earlier in this chapter is a simple example of unsupervised learning. Transfer learning (TL) is a learning framework that enables knowledge transferring among task domains. In practice this means that the method enables fast and less data-greedy learning of the signal environment by using data collected for other tasks. Such a capability is valuable especially in Wi-Fi fingerprinting due to its requirement for laborious offline training. TL framework was built in [42] to capture the distance metric presented in Section 3.4 from environments with sufficient training data. Then, a selection operator is designed to identify the appropriate knowledge from the learned metrics for the target domain by minimizing the data discrepancy between target and source domains. The framework was evaluated to reduce offline training time and enhance system scalability.

3.8 Summary

In this chapter we studied the methods for how the various measurements presented in Chapter 2 can be leveraged for the estimation of position, velocity, and/or orientation. As measurements are inevitably uncertain to some extent, it is important to understand the propagation of measurement errors into position estimation uncertainty, which is often represented in terms of statistics. For static estimation problems, the CRB gives the best possible error covariance that can be (at least theoretically) achieved.

Especially when there are redundant measurements, closed-form equations are often inconvenient for estimation. If the system of equations is linear and the measurement error variances are known, the least squares give the minimum-variance solution among linear estimators. In practice, the measurement models are often nonlinear and a more advanced algorithm (e.g., iterative Gauss–Newton) is needed for static least-squares estimation. Nevertheless, cases exist where least-squares estimation does not work and the uncertainty is difficult to quantify; fingerprinting is an example.

A fundamental property of most positioning problems is that the position (as well as velocity and orientation, if applicable) estimates are highly correlated in time, and therefore, it makes sense to estimate them as a time series. In practice, this needs a model how the unknowns (i.e., the state variables of the system) evolve in time. If both the state evolution model and the measurement models are linear with Gaussian uncertainty, the KF gives the optimal solution in terms of estimation uncertainty variance. However, the nonlinear EKF is often needed in practice, and the optimality can no longer be guaranteed. In most challenging scenarios, one can resort to computationally expensive Monte Carlo methods such as the PF that can work with any measurement model or error distribution.

Emerging trends in the field of positioning algorithms include factor graph optimization and machine learning. In factor graph optimization, the dependencies of the state-space are modeled as a graph, and the entire time series of states is determined at once by optimization; nevertheless, it has been shown that the solution can be computed incrementally over time instead of repeated batch solutions. Moreover, ML techniques such as deep and transfer learning show great promise in tackling practical challenges such as multisensor synchronization and calibration imperfections as well as complex measurement error models.

In the following chapter we will extend the scope to the navigation system level that includes aspects such as maps and multiple collaborative users. We will also study the performance of a radio-based indoor navigation system that utilizes an EKF for position estimation.

References

[1] European Telecommunications Standards Institute (ETSI), *TR 138 901, 5G; Study on Channel Model for Frequencies from 0.5 to 100 GHz*, 2017.

[2] Savage, P. G., "Computational Elements for Strapdown Systems," RTO Educational Notes RTO-SET-116, 2008, NATO Research and Technology Organization, 2009.

[3] Kay, S. M., *Fundamentals of Statistical Signal Processing, Volume I: Estimation Theory*, Prentice Hall, 1993.

[4] Koivisto, M., M. Costa, J. Werner, et al., "Joint Device Positioning and Clock Synchronization in 5G Ultra-Dense Networks," *IEEE Transactions on Wireless Communications*, Vol. 16, No. 5, 2017, pp. 2866–2881.

[5] Sand, S., A. Dammann, and C. Mensing, *Positioning in Wireless Communications Systems*, Chichester, U.K.: Wiley, 2014.

[6] Werner, J., J. Wang, A. Hakkarainen, D. Cabric, and M. Valkama, "Performance and Cramer-Rao Bounds for DoA/RSS Estimation and Transmitter Localization Using Sectorized Antennas," *IEEE Transactions on Vehicular Technology*, Vol. 65, No. 5, 2016, pp. 3255–3270.

[7] Abu-Shaban, Z., X. Zhou, T. Abhayapala, G. Seco-Granados, and H. Wymeersch, "Error Bounds for Uplink and Downlink 3D Localization in 5G Millimeter Wave Systems," *IEEE Transactions on Wireless Communications*, Vol. 17, No. 8, 2018, pp. 4939–4954.

[8] Shahmansoori, A., G. E. Garcia, G. Destino, G. Seco-Granados, and H. Wymeersch, "Position and Orientation Estimation Through Millimeter-Wave MIMO in 5G Systems," *IEEE Transactions on Wireless Communications*, Vol. 17, No. 3, 2018, pp. 1822–1835.

[9] Blum, A., J. Hopcroft, and R. Kannan, *Foundations of Data Science*, Cambridge, U.K.: Cambridge University Press, 2020.

[10] Kaplan, E., and D. Hegarty, *Understanding GPS: Principles and Applications*, Norwood, MA: Artech House, 2006.

[11] Gratton, S., A. S. Lawless, and N. K. Nichols, "Approximate Gauss—Newton Methods for Nonlinear Least Squares Problems," *SIAM Journal on Optimization*, Vol. 18, No. 1, 2007, pp. 106–132.

[12] IndoorAtlas, "Build Seamless Location-Based Experiences Today," September 2022, https://www.indooratlas.com.

[13] Pei, L., J. Liu, R. Guinness, Y. Chen, T. Kroger, R. Chen, and L. Chen, "The Evaluation of WiFi Positioning in a Bluetooth and Wifi Coexistence Environment," in *2012 Ubiquitous Positioning, Indoor Navigation, and Location Based Service (UPINLBS)*, 2012, pp. 1–6.

[14] Pei, L., R. Chen, J. Liu, H. Kuusniemi, T. Tenhunen, and Y. Chen, "Using Inquiry-Based Bluetooth RSSI Probability Distributionsfor Indoor Positioning," *Journal of Global Positioning Systems*, Vol. 9, No. 2, 2010, pp. 122–130.

[15] Groves, P. D., *Principles of GNSS, Inertial, and Multisensor Integrated Navigation Systems*, Second Edition, Norwood, MA: Artech House, 2013.

[16] Moreira, A., M. J. Nicolau, F. Meneses, and A. Costa, "Wi-Fi Fingerprinting in the Real World–RTLS@UM at the EvAAL Competition," in *2015 International Conference on Indoor Positioning and Indoor Navigation (IPIN)*, 2015, pp. 1–10.

[17] Shkel, A. M., "Inertial MEMS Sensors Are Becoming 3D and Atomically Precise," Keynote, in *Proceedings of the International Conference on Indoor Positioning and Indoor Navigation (IPIN 2018)*, Nantes, France, September 2018.

[18] Skog, I., P. Handel, J.-O. Nilsson, and J. Rantakokko, "Zero-Velocity Detection: An Algorithm Evaluation," *IEEE Transactions on Biomedical Engineering*, Vol. 57, 2010, 2657–2666.

[19] Foxlin, E., "Pedestrian Tracking with Shoe-Mounted Inertial Sensors," *IEEE Computer Graphics and Applications*, Vol. 25, No. 6, 2005, pp. 38–46.

[20] Ruotsalainen, L., M. Kirkko-Jaakkola, J. Rantanen, and M. Makela, "Error Modelling for Multi-Sensor Measurements in Infrastructure-Free Indoor Navigation," *Sensors*, Vol. 18, No. 2, 2018.

[21] Schepers, H. M., H. F. J. M. Koopman, and P. H. Veltink, "Ambulatory Assessment of Ankle and Foot Dynamics," *IEEE Transactions on Biomedical Engineering*, Vol. 54, No. 5, 2007, pp. 895–902.

[22] Zhang, R., F. Hoflinger, and L. Reindl, "Inertial Sensor Based Indoor Localization and Monitoring System for Emergency Responders," *IEEE Sensors Journal*, Vol. 13, No. 2, 2013, pp. 838–848.

[23] Walder, U., and T. Bernoulli, "Context-Adaptive Algorithms to Improve Indoor Positioning with Inertial Sensors," in *International Conference on Indoor Positioning and Indoor Navigation (IPIN)*, 2010, pp. 1–6.

[24] Park, S. K., and Y. S. Suh, "A Zero Velocity Detection Algorithm Using Inertial Sensors for Pedestrian Navigation Systems," *Sensors*, Vol. 10, No. 10, 2010, pp. 9163–9178.

[25] Wagstaff, B., V. Peretroukhin, and J. Kelly, "Improving Foot-Mounted Inertial Navigation Through Real-Time Motion Classification," in *International Conference on Indoor Positioning and Indoor Navigation (IPIN)*, 2017, pp. 1–8.

[26] Rantanen, J., M. Makela, L. Ruotsalainen, and M. Kirkko-Jaakkola, "Motion Context Adaptive Fusion of Inertial and Visual Pedestrian Navigation," in *2018 International Conference on Indoor Positioning and Indoor Navigation (IPIN)*, IEEE, 2018, pp. 206–212.

[27] Hermann, R., and A. J. Krener, "Nonlinear Controllability and Observability," *IEEE Transactions on Automatic Control*, Vol. AC-22, No. 5, October 1977, pp. 728–740.

[28] Kalman, R. E., "A New Approach to Linear Filtering and Prediction Problems," *Journal of Basic Engineering*, Vol. 82, No. 1, 1960, pp. 35–45.

[29] Sarkka, S., *Bayesian Filtering and Smoothing*, Cambridge, U.K.: Cambridge University Press, 2013.

[30] Mehra, R. K., and J. Peschon, "An Innovations Approach to Fault Detection and Diagnosis in Dynamic Systems," *Automatica*, Vol. 7, No. 5, 1971, pp. 637–640.

[31] Pitt, M. K., and N. Shephard, "Filtering Via Simulation: Auxiliary Particle Filters," *Journal of the American Statistical Association*, Vol. 94, No. 446, 1999, pp. 590–599.

[32] Kitagawa, G., "Monte Carlo Filter and Smoother for Non-Gaussian Nonlinear State Space Models," *Journal of Computational and Graphical Statistics*, Vol. 5, No. 1, 1996, pp. 1–25.

[33] Del Moral, P., A. Doucet, and A. Jasra, "On Adaptive Resampling Strategies for Sequential Monte Carlo Methods, *Bernoulli*, Vol. 18, No. 1, February 2012, p. 252–278.

[34] Dellaert, F., and M. Kaess, "Factor Graphs for Robot Perception, *Foundations and Trends in Robotics*, Vol. 6, Nos. 1–2, 2017, pp. 1–139.

[35] Loeliger, H.-A., "An Introduction to Factor Graphs," *IEEE Signal Processing Magazine*, Vol. 21, No. 1, January 2004, pp. 28–41.

[36] Kaess, M., H. Johannsson, R. Roberts, V. Ila, J. J. Leonard, and F. Dellaert, "iSAM2: Incremental Smoothing and Mapping Using the Bayes Tree," *The International Journal of Robotics Research*, Vol. 31, No. 2, 2012, pp. 216–235.

[37] Joswig, N., J. Autiosalo, and L. Ruotsalainen, "Improved Deep Depth Estimation for Environments with Sparse Visual Cues," *Machine Vision and Applications*, Vol. 34, No. 18, 2023.

[38] Mnih, V., K. Kavukcuoglu, D. Silver, A. Graves, I. Antonoglou, D. Wierstra, and M. Riedmiller, "Playing Atari with Deep Reinforcement Learning," *NEURIPS*, 2013.

[39] Abid, M., P. Compagnon, and G. Lefebvre, "Improved CNN-Based Magnetic Indoor Positioning System Using Attention Mechanism," in *2021 International Conference on Indoor Positioning and Indoor Navigation (IPIN)*, 2021, pp. 1–8.

[40] Liu, L., G. Li, and T. H. Li, "Atvio: Attention Guided Visual-Inertial Odometry," in *ICASSP 2021–2021 IEEE International Conference on Acoustics, Speech and Signal Processing (ICASSP)*, 2021, pp. 4125–4129.

[41] Sheikhpour, S., and M. Maher Atia, "Calibration-Free Visual Inertial Fusion with Deep Convolutional Recurrent Neural Networks," in *Proceedings of the 32nd International Technical*

Meeting of the Satellite Division of The Institute of Navigation (ION GNSS+ 2019), 2019, pp. 2198–2209.

[42] Liu, K., H. Zhang, J. Kee-Yin Ng, Y. Xia, L. Feng, V. C. S. Lee, and S. H. Son, "Toward Low-Overhead Fingerprint-Based Indoor Localization Via Transfer Learning: Design, Implementation, and Evaluation," *IEEE Transactions on Industrial Informatics*, Vol. 14, No. 3, 2018, pp. 898–908.

[43] Sutera, E., V. Mazzia, F. Salvetti, G. Fantin, and M. Chiaberge, "Indoor Point-to-Point Navigation with Deep Reinforcement Learning and Ultra-Wideband," in *Proceedings of the 13th International Conference on Agents and Artificial Intelligence - Volume 1: ICAART*, SciTech Press, 2021, pp. 38–47.

[44] Clark, R., S. Wang, H. Wen, A. Markham, and N. Trigoni, "Vinet: Visual-Inertial Odometry as a Sequence-to-Sequence LearningProblem," *Proceedings of the AAAI Conference on Artificial Intelligence*, Vol. 31, No. 1, 2017.

[45] Kim, K. S., S. Lee, and K. Huang, "A Scalable Deep Neural Network Architecture for Multi-Building and Multi-Floor Indoor LocalizationBased On Wi-Fi Fingerprinting," *Big Data Analytics*, Vol. 3, No. 4, 2017.

4

Navigation System Setup

Up to now we have discussed the measurements used for indoor navigation and algorithms turning the measurements into a positioning or navigation solution. However, coordinates forming the navigation solution are not enough if their presentation is not in a form that can be understood by humans. This chapter begins in Section 4.1 by looking at methods connecting the navigation solution with a map for both representation purposes and then for improving the obtained solution. When navigation needs to be done infrastructure-free (i.e., without any fixed system preparation) SLAM provides a full navigation solution. SLAM is then covered in Section 4.2. In situations where an accurate and reliable navigation solution is needed for a group of users in unknown indoor environments, cooperative navigation should be considered. Cooperative navigation enables sharing of information between users, thereby improving everyone's individual navigation solution. Cooperative navigation will be discussed in Section 4.3. In Section 4.4, we will change the viewpoint a bit and discuss tracking users in the environment. As the navigation using radio signals is mainly implemented as tracking, meaning that the position is computed outside of the user device, here our discussion concentrates on computer vision based tracking. We will finish the chapter by presenting a full indoor radio navigation system and providing representative examples with code to be used and that can be developed further for the reader's own use.

4.1 Maps

In many positioning applications, the resolved position will ultimately be visualized on a preexisting map. GNSS-based car navigation systems are a common example in the outdoor context, but it is easy to see that plain coordinates without a map are rarely useful in indoor positioning applications such as robotics, asset management, or first responder tracking.

Often, map information can be leveraged as a constraint in the position estimation for improved performance, which is known as *map matching* [1]. In the examples mentioned above, the motion of a car is in most cases constrained to the road network, and in indoor applications, the time series of position estimates should be consistent with the room layout. The effect of map matching depends significantly on the topology of the building: map constraints can yield a good navigation performance with relatively inaccurate navigation sensors in an office building consisting of small rooms and narrow hallways, but this is not the case in an exhibit hall.

When implementing map matching, the obvious first question is how to obtain the map. If the application is limited to a certain building (e.g., asset management in a warehouse), a detailed map is often readily available. For more generic applications like personal navigation assistance, services like Google Maps include floor plans of many large venues such as shopping malls and airports. For use cases where a priori mapping is simply not available, even the possibility of creating maps on the fly by photographing fire evacuation maps has been studied [2]; in this approach, an obvious challenge is to determine the scale of the map. Another source of coarse map information could be an aerial photograph, which obviously only shows the layout of the exterior walls. Alternatively, one may utilize a SLAM solution instead of map matching. In [3], an explicit map of the venue was not used, but the building was *assumed* to have a grid-like layout, and the heading was constrained to follow the cardinal directions of the grid. However, as acknowledged in [3], the assumption is not applicable to all buildings.

4.1.1 Map Matching with Particle Filter

Perhaps the most intuitive approach to indoor map matching is to enforce a constraint that the position estimate must not cross any interior or exterior walls. In the context of Bayesian filtering, this map constraint corresponds to a measurement likelihood function of the type

$$P\left(\text{map}|\boldsymbol{x}_t, \boldsymbol{x}_{t-1}\right) = \begin{cases} 0 & \text{if there is a wall between positions } \boldsymbol{x}_{t-1} \text{ and } \boldsymbol{x}_t \\ 1 & \text{otherwise} \end{cases}. \tag{4.1}$$

It is obvious that this highly nonlinear (and nondifferentiable) likelihood function cannot be applied to Kalman-type filters. In contrast, (4.1) is straightforward to use in a PF; essentially, it implies that particles that collide with walls are discarded.

Applying a likelihood similar to (4.1) has been extensively studied in literature. The use of body-mounted PDR together with a 2D map was investigated in [4]. However, such an approach is not well suited to multifloor buildings; the system should be augmented with, for example, a barometer to detect floor changes. In [5], a foot-mounted IMU was used together with a detailed 3D map to accommodate the height coordinate.

In principle, using a PF with position and heading states to combine PDR and map matching may yield a satisfactory performance for many indoor positioning applications. The concept is illustrated in Figure 4.1. However,

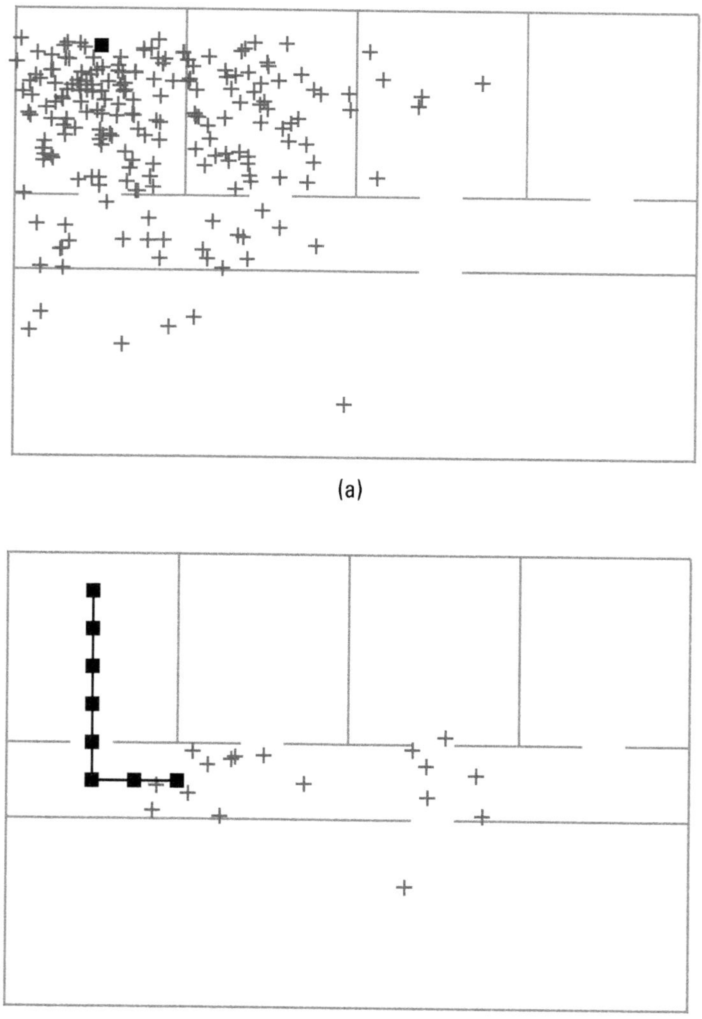

Figure 4.1 Particle filter with wall crossing constraints. Particles are represented by crosses (discarded ones are not shown), and solid squares denote the true route. Resampling is not applied. (a) Initial particles, and (b) particles remaining after motion.

the approach has several practical challenges. First, the algorithm needs to be initialized such that the true initial position and heading is included in the cloud of particles. If there is no other means of positioning (e.g., Wi-Fi), to bound the uncertainty of the initial state, covering the entire building area with sufficient heading diversity may require too many particles and become impractical. Furthermore, if the building layout is considerably symmetric, the cloud of particles may remain multimodal—this is the case in Figure 4.1(b) where a group of particles initially located in the room adjacent to the true one remains alive. Second, the measurement model (4.1) is sensitive to errors in the map information: if the user traverses a passage where the map shows (a nonexistent) wall, particles will be discarded unnecessarily. In practice, it may be necessary to use a small positive value for the measurement function when crossing walls instead of 0 in (4.1) to prevent the loss of all particles. Alternative solutions include artificially inflating the variance of the cloud of particles to occasionally allow some particles to penetrate walls.

If the initial position uncertainty is large or the motion sensors are inaccurate, the wall crossing constraint (4.1) often discards a large amount of particles, which leads to sample impoverishment. The situation is illustrated in Figure 4.1(b) where resampling has been omitted to show much information from the initial set of particles has been retained; in practice, one would resample the particles well before their number falls this low. Another property of map constraints is that they can make the particle distribution very nonsymmetric, such as in Figure 4.1(a): when the user lies close to a wall, most of the particles can be located on only one side of it, which causes the (weighted) mean of the particles to fall further away from the wall.

4.1.2 Graph-Based Map Constraints

The relatively good performance of the map-matching particle filter is obtained at the price of higher computational complexity as well as the requirement of detailed map information. The computational efficiency of map matching can be improved by processing the map information in the form of a graph instead of a set of wall segments.

A popular map representation is the Voronoi graph. The vertices of a Voronoi graph are chosen such that the cells of the corresponding Voronoi diagram correspond to the floor plan; the Voronoi cell of vertex i is the region where the distance to any other vertex is no shorter than the distance to vertex i. An example Voronoi diagram is shown in Figure 4.2(a). The edges of a Voronoi graph simply define the possible transitions from one vertex to another; a Voronoi graph representation of a simple indoor map is plotted in Figure 4.2(b). Given a floor plan in the form of an image, a Voronoi graph can be automatically generated [6].

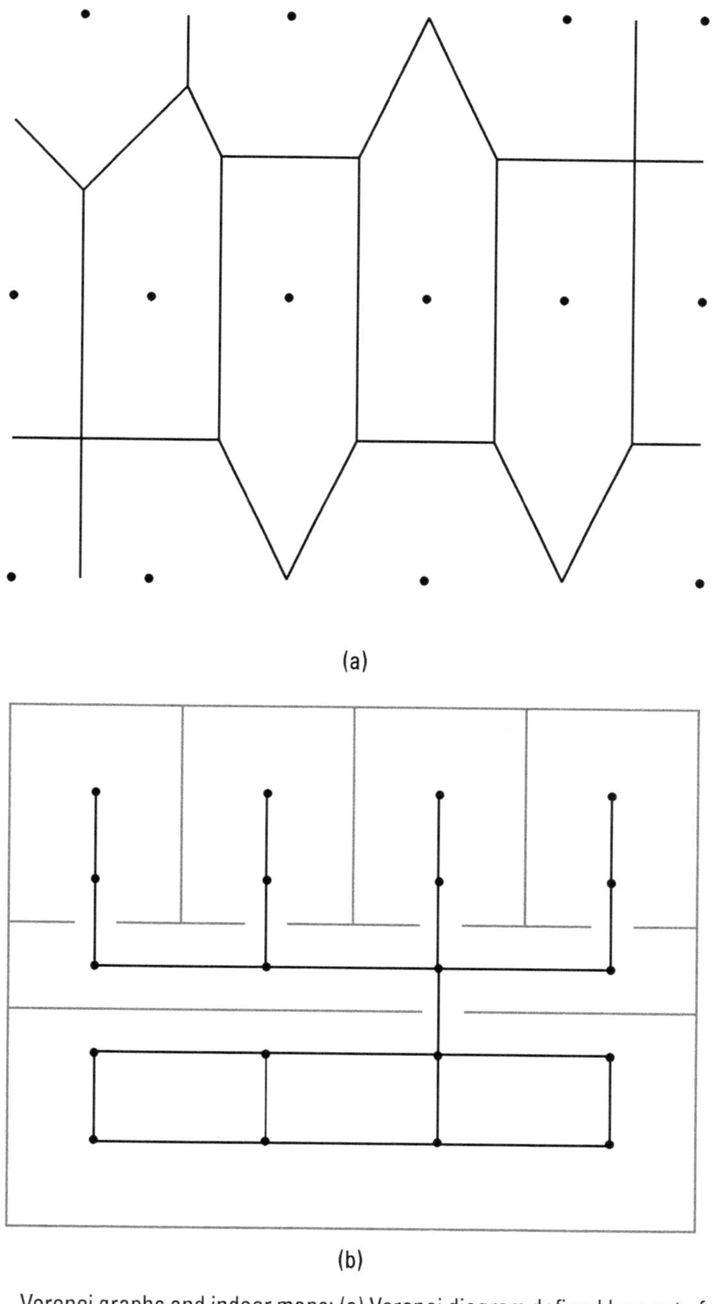

Figure 4.2 Voronoi graphs and indoor maps: (a) Voronoi diagram defined by a set of 14 points, and (b) indoor map and a corresponding Voronoi graph.

A simple approach is to constrain the user position onto the vertices and edges of the map graph. For instance, the state vector can consist of three components: an edge identifier, the distance from the starting vertex of the edge (assuming the graph is directed), and a Boolean variable to indicate whether the user is moving or stationary [7]. This model is still unsuitable for a Kalman-type filter, but a PF implementation does not need as many particles as the approach based on (4.1) [6]. It is also possible to design a state model that allows using another estimation algorithm instead of PF, such as, the Viterbi algorithm [1]. The obvious drawback of constraining the user position onto a graph is that the model will not include all possible user positions, which inherently limits the attainable accuracy in comparison with the wall-crossing constraint. The coverage of the graph model can be improved by increasing the number of nodes or complementing the model with open spaces (i.e., regions inside which the user position is unconstrained) [3].

Grid representations are another popular approach to graph-type map matching [9]; the principle is illustrated in Figure 4.3. A grid map discretizes the position state space, and the likelihood of the user position is maintained for each grid cell. When the grid is derived from a map, certain cells, such as those containing walls, can be constrained to zero probability of the user occupying them. The granularity of the grid can be denser than 1m—for instance, at the level of the average step length of a human for pedestrian applications [10]—but it is obvious that increasing the grid density also increases the computational complexity. In practice, the discrete state space quickly becomes so large that updating the likelihood for all cells with each received measurement does not

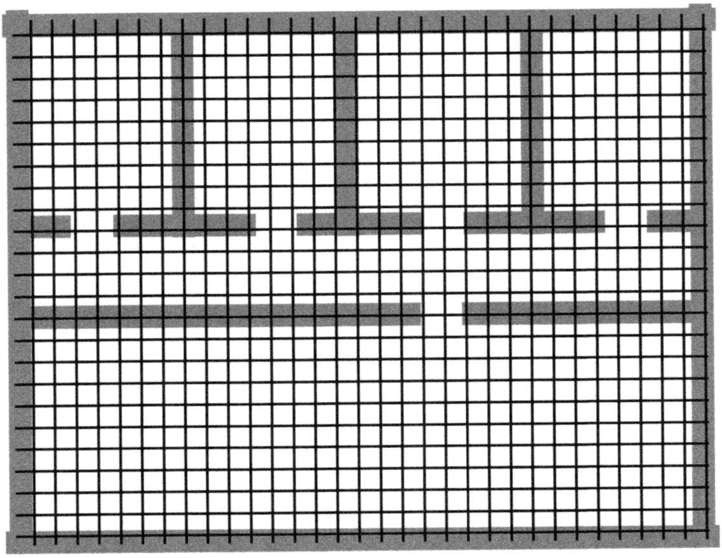

Figure 4.3 Indoor map represented by a grid.

make sense; logic to select only the relevant cells for measurement updates is necessary [9, 11].

4.2 Simultaneous Localization and Mapping

SLAM is a key technology for providing accurate and reliable infrastructure-free indoor navigation solution. In SLAM, the user simultaneously creates a map of the environment in order to use this map for positioning with the help of probability models. The technique providing accuracy and reliability into this relative navigation method is called loop closure. Loop closure happens when the user returns to a previously visited location. The revisit is detected by finding a match between the present location with the information about previously visited locations in a database. At loop-closure drifting location errors can be corrected and the accuracy of the map can be improved.

SLAM does not need any a priori knowledge of the location or environment. Initially SLAM was known as probabilistic mapping and formulated as the simultaneous estimation of 3D user and landmark poses. We will start the discussion about SLAM by presenting the probabilistic SLAM setting due to its clarity and relevance to the topic and will present the state-of-the-art methods later.

4.2.1 Probabilistic SLAM

The user pose, location, and orientation, is represented with a state vector $\mathbf{x}_k$ at time instant k. It is computed based on the measurements of the landmark locations i with respect to the user obtained with one or multiple sensors [12]. Measurements are obtained using a camera, lidar, inertial sensors, or any other equipment enabling observations ($\mathbf{z}_{ki}$) of the landmarks' locations and orientations with respect to the user. Landmarks ($\mathbf{l}_i$) are (x, y)-points in the 2D world, or respectively (x, y, z)-points in 3D. The world dimensions are defined by the sensor used; for example, 2D for a monocular camera and 3D for lidar. Figure 4.4 shows the SLAM setting. In addition to the state, observation, and landmark vectors, the SLAM algorithm includes a control vector, $\mathbf{u}_k$, which controls the transition of the system between the states.

Probabilistic SLAM algorithm computes the conditional probability distribution

$$P(\mathbf{x}_k, \mathbf{l} | \mathbf{z}_{0:k}, \mathbf{u}_{0:k}, \mathbf{x}_0) \tag{4.2}$$

at each time step k. The notation $0{:}k$ means that information about the previous measurements and control input will be incorporated. This joint posterior density of the landmark locations and user's state, conditional to the observations, control input, and initial state, may be computed using Bayes's theorem. The computation requires three mathematical models: motion, inverse, and direct observation.

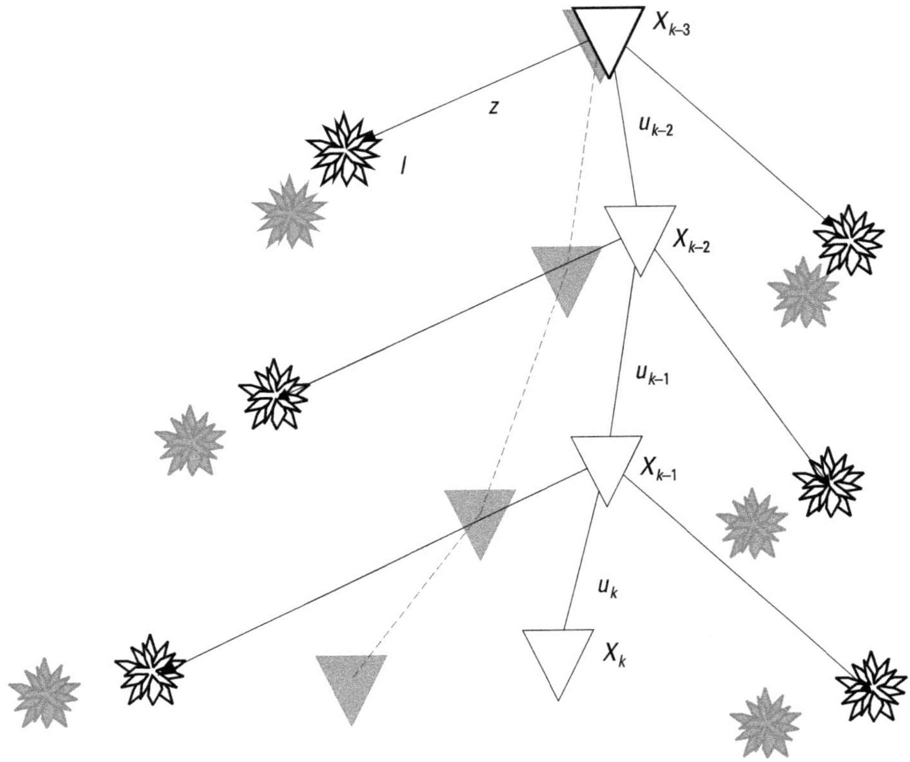

Figure 4.4 Probabilistic SLAM principle. $\mathbf{x}_k$ are state vectors at time instants k, $\mathbf{z}$ are measurements, $\mathbf{l}$ landmarks, and $\mathbf{u}_k$ control vectors.

The motion model gives mathematical representation for the user's motion, a Markov process describing the state $\mathbf{x}_k$, which is independent of the observations and landmarks and depends only on the previous state $\mathbf{x}_{k-1}$. It is described by a probability distribution on the state transition as $P(\mathbf{x}_k|\mathbf{x}_{k-1}, \mathbf{u}_k)$.

Observation models are related to landmark sensor measurements. Inverse observation model estimates the uncertain measurements of landmarks and incorporates those to the map using the uncertain user's localization solution. Then, the direct observation model is used to decrease the uncertainty in user and landmark localization solutions via incorporating previously mapped landmarks. Process provides the probability of making an observation $\mathbf{z}_k$ with the estimated user state $\mathbf{x}_k$ and landmark location $\mathbf{l}$ as $P(\mathbf{z}_k|\mathbf{x}_k, \mathbf{l})$.

Solving the a posteriori distribution (4.2) is difficult because it contains dependencies; the persons to be located need a map to locate themselves, but at the same time, positioning is needed to create a map. Fortunately, the problem can be solved iteratively by solving the user's location and position relative to the starting point, and this information can be further used to complete the map. At its simplest, the created map can be just a line showing the route taken, and at its richest, a 3D model of the environment.

Probabilistic SLAM solves the localization and mapping task in two recursive steps; by computing the prior distribution in prediction (or time-update) step where the state is updated using previous states and measurements (4.3), and by computing the posterior distribution in the correction (or measurement update) when measurements are obtained (4.4) as

$$P(\mathbf{x}_k, \mathbf{l}|\mathbf{z}_{0:k-1}, \mathbf{u}_{0:k}, \mathbf{x}_0) = \int P(\mathbf{x}_k, |\mathbf{x}_{k-1}, \mathbf{u}_k) \times P(\mathbf{x}_{k-1}, \mathbf{l}|\mathbf{z}_{0:k-1}, \mathbf{u}_{0:k-1}, \mathbf{x}_0) d\mathbf{x}_{k-1} \quad (4.3)$$

for the prior distribution and

$$P(\mathbf{x}_k, \mathbf{l}|\mathbf{z}_{0:k}, \mathbf{u}_{0:k}, \mathbf{x}_0) = \frac{P(\mathbf{z}_k|\mathbf{x}_k, \mathbf{l}) P(\mathbf{x}_k, \mathbf{l}|\mathbf{z}_{0:k-1}, \mathbf{u}_{0:k}, \mathbf{x}_0)}{P(\mathbf{z}_k|\mathbf{z}_{o:k-1}, \mathbf{u}_{0:k})} \quad (4.4)$$

for the posterior distribution. Therefore, solving the probabilistic SLAM problem involves finding an appropriate representation for both the observation and motion models and thereby the prior and posterior distributions. If the location of the user $\mathbf{x}_k$ is known for all time steps, the probabilistic algorithm may be degenerated into solving the landmark map only as $P(\mathbf{l}|\mathbf{x}_{0:k}, \mathbf{z}_{0:k}, \mathbf{u}_{0:k})$. Similarly, if the landmark locations (the map) are known, user localization solution may be computed from $P(\mathbf{x}_k|, \mathbf{z}_{0:k}, \mathbf{u}_{0:k}, l)$. By considering the recursive nature of the process and looking at (4.3) and (4.4), it is not a surprise that previously SLAM was solved mainly using the EKF and particle filtering discussed in Chapter 3. At present, more advanced methods are used and will be discussed in the next sections.

In the following, we will first look at purely visual SLAM, and then SLAM processed with other measurements, such as ones obtained using IMUs or radio positioning means.

4.2.2 Visual SLAM

Visual SLAM (vSLAM) has already been an active research area for decades, first concentrating on SLAM techniques for robots and then recently moving to solutions for pedestrians, and therefore inherently advancing the algorithms to function with a monocular camera. Despite the growing maturity of the methods, state-of-the-art vSLAM methods still, at the time of writing, struggle to simultaneously achieve accuracy, robustness, and real-time capabilities. This struggle has directed research activities toward deep learning based methods. However, the SLAM methods used in safety-critical applications, for example industrial ones, will most likely be based on more traditional methods far into the future due to the challenges of understanding the decision-making principles of deep learning methods.

The VSLAM pipeline consists of the following five steps: (1) initialization, (2) localization based on computing the motion between consecutive images similarly as in visual odometry, (3) mapping, (4) global optimization, namely loop-closure and a refining map based on that, and (5) relocalization (i.e., restarting the SLAM process with previously saved map or when location is lost during the process).

VSLAM solutions are divided into two classes: filter-based, which are those using Kalman or particle filtering as discussed previously, and keyframe-based, which are those relying on optimization to estimate both localization and mapping of the environment. As described above for probabilistic SLAM, filter-based systems process each image and recursively update both the location and landmark-based map. Since 2010 [13] keyframe-based methods have been dominating the field due to their improved performance. In keyframe-based systems, seven steps are used. First, during *visual initialization*, an initial map and user state are established. Then, in *data association*, the previous state is used to predict the present state (*pose estimation*) and to establish associations with the map (*map generation*). The residual (i.e., the difference between the predicted and actual states) is iteratively minimized and if it diverges or the data association fails, *failure recovery* process is started. The method is called keyframe-based, as some of the images are selected as keyframes and used for expanding the map in a process called *bundle adjustment*. Bundle adjustment is a process of simultaneously adjusting the pose and landmark (3D point) locations for multiple consecutive overlapping images by minimizing a global cost function. This nonlinear minimization of the cost function, namely the measurement reprojection errors, requires nonlinear least-squares optimization and is generally solved using the Levenberg-Marquart algorithm. At the same time as this map optimization is done, the loop closures are detected to correct the estimated location and map.

SLAM may be divided into four classes based on the processing technique: two for localization (indirect and direct), and two for mapping (sparse and dense). The indirect SLAM methods are based on detecting and tracking image features such as the previously discussed SIFT features or at present generally their more advanced variants ORBs. Indirect SLAM methods are more robust to photometric variations than their direct counterparts. Direct VSLAM uses all image pixels directly by using brightness transfer function parameter. The strengths of indirect SLAM compared to indirect are its fastness, flexibility in removing outliers from observations, and robustness to inconsistencies in the models. Also, indirect SLAM does not need good initialization. Direct SLAM from its behalf, although it fails in all previously mentioned characteristics, can reconstruct the whole scene seen in images and thereby its decisions are based on complete information. Terms sparse and dense are related to the mapping process. Sparse maps include only the detected features, such as corners and other local representations. They don't have any notions of neighborhoods. Dense maps include all pixels from images used for the process and therefore they exploit the connectedness of the image regions used. Figure 4.5 shows a snapshot from a dense point cloud map formed with an iPhone model 13's lidar. Contrary to a popular belief, indirect and sparse or direct and dense are not always simultaneous, but all combinations of types exist. Table 4.1 shows the most well-known implementations of each of the four classes. At the time of writing, SLAM methods from the ORB-SLAM family were considered as the most popular state-of-the-art SLAMs due to their accuracy, processing speed, and process transparency. ORB-SLAM as well as other state-of-the-art methods will be discussed in Section 4.2.2.2.

SLAM systems include two main components: the front-end and the back-end. The front-end extracts relevant features from the images. The front-end also associates the features with the corresponding landmarks, namely, it does the data association. Finally, the front-end provides an initial guess for the variables in the nonlinear optimization. The back-end is responsible for the global optimization of an accurate SLAM solution.

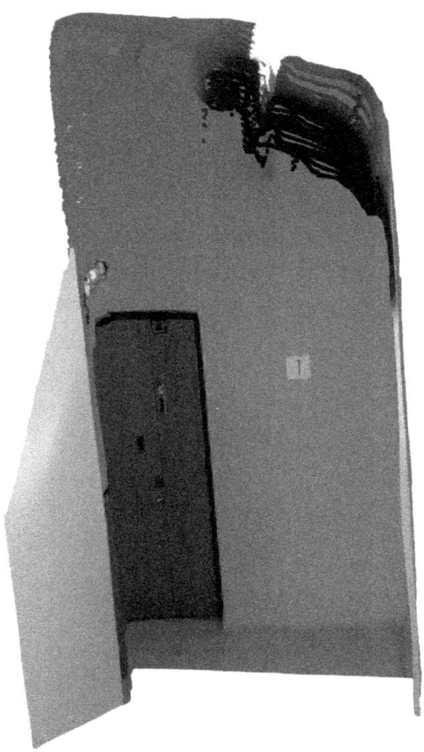

Figure 4.5 Dense SLAM point cloud map done using iPhone 13's lidar.

Table 4.1
Four Classes of SLAM

Processing Class	Principle	Example Implementation
Sparse + Indirect	Estimates 3D geometry from a set of keypoint matches	**ORB-SLAM2**
Dense + Indirect	Estimates 3D geometry from a dense, regularized optical flow field	**DENSEMD**
Sparse + Direct	Optimizes a photometric error defined directly on the images, no geometric prior	**DSO**
Dense + Direct	Employs a photometric error as well as a geometric prior	**LSD-SLAM**

Fortunately, many of the state-of-the-art SLAM algorithms are published as open-source code. References for the repository links, with classification related to the previous discussion about dense/sparse and direct/indirect SLAM are given in Table 4.2.

In order for the new visual measurements to be matched to the previous ones, the measurements must overlap. The search space should be limited if possible, for example by making assumptions about possible directions and speeds. Otherwise, the search space can be huge, especially with 3D measurements. If the movement is too fast in relation to the measurement frequency, the method may fail completely. However, the situation is made easier if SLAM is done with the help of several sensors. In this case, for example, an inertial unit with a high measurement frequency provides valuable additional information to support vision's slow image processing. Moving objects in the camera's path, such as pedestrians or vehicles, can also mess up the SLAM solution. When the camera follows a moving object, the displacement obtained no longer describes the displacement of the camera, but the sum of the displacements of the camera and the moving object. However, even this problem can be circumvented with a fusion of several sensors. In a setting where the user motion is obtained from control inputs—that is, measurements of sensors u_i (for example, an IMU or an odometer), and environmental information

Table 4.2
Open-Source Visual SLAM Architectures

Name	Localization	Mapping	Setup	References
ORB-SLAM2	Indirect	Sparse	Monocular stereo RGB-D	[14]
DENSEMD	Indirect	Dense	Monocular	[15]
LSD-SLAM	Direct	Dense	Monocular	[16]
DSO	Direct	Sparse	Monocular	[17]

z_t produced by imaging—the former are considered internal (proprioceptive) and the latter external (exteroceptive) measurements.

4.2.2.1 Loop Closure

As mentioned previously, loop closure is the mechanism that improves the accuracy of the computed location and map and therefore the reliability of the SLAM by correcting the drifting relative solutions. However, an erroneous loop closure might destroy the solution and therefore the aim is to obtain zero false positives in detection; namely, not detecting a nonexisting loop.

The first step in loop closure is to detect that the place seen in the images captured during the process has been visited before. When the navigation area is large, loops occur sparsely or even never. Fortunately, an indoor navigation setting restricts the area, and possibilities for committing a loop closure are much higher than outdoors. While the user navigates and observes the environment, new observations are added to the database from each image taken. The observations are features detected from images, either global or local, as discussed in Chapter 3, or more and more via deep learning. Using the extracted features, the image sequence gathered is described by a database of visual representations. Loop closure detection pipelines are divided into single-image and sequence-of-images-based. Methods in the first category look for the most identical view from the database, while methods in the second category look for the matching location between submaps (i.e., groups of individual images). To match the features detected from the present image with the ones in the database, the images immediately preceding should not be used. The reason for excluding the immediately preceding images is that their features are usually similar in appearance to the present view although the area is not revisited. While searching for matches from the database, a sliding window defined either by a timing constant or environmental semantic changes is used for rejecting such features. To decide whether a loop occurs, a similarity score between the present (query) and features from any previously seen images is computed and compared against a similarity threshold. In the case of single-image methods an individual similarity score for each database entry is computed and for sequence-to-sequence to the submaps.

Features and their representations are crucial for the success of SLAM. The dynamic objects accidentally included into the representations and changing environmental factor, such as illumination and seasonal changes, complicate the task, especially outdoors. Loop closure should ideally achieve performance indicated with the recall at 100% precision (RP100) measure despite the navigation environment. It represents the highest possible recall score for perfect precision, meaning that there are no false-positives in loop-closure detection. A single false-positive loop-closure detection can cause a total failure for SLAM and therefore RP100 is a good indicator. Precision and recall are computed from the true-positive, false-positive, and false-negative loop-closure detection results as

$$\text{Precision} = \frac{\text{True-positives}}{\text{True-positives} + \text{False-positives}} \tag{4.5}$$

$$\mathit{Recall} = \frac{\text{True-positive}}{\text{True-positive} + \text{False-negatives}} \tag{4.6}$$

The closer to 1 the value of recall is for the precision fixed by definition to 1, the better the performance of the loop-closure detection is. After detection, the predicted location and map must be corrected. Pose-graph optimization and bundle adjustment (BA) have been used to reduce the accumulated error. Pose-graph optimization optimizes the error by building a pose graph with two camera poses connected when they see the same feature and Bundle adjustment minimizes the error over 3D features and camera poses simultaneously [18]. Figure 4.6 shows a simple illustration of the effect of loop closure for the localization. In Figure 4.6(a), the solution is obtained without loop-closure drifts as the path remains too short due to underestimated translation and errors in orientation. In Figure 4.6(b), algorithm has detected a revisit and corrected the location.

4.2.2.2 Future of Visual SLAM

The eternal challenge in monocular SLAM's performance is the depth issue discussed before. The depth, or distance between the user and landmark, creates the unknown scale in motion and mapping and must therefore be solved to obtain an accurate solution. The early SLAM algorithms used ultrasound or other range-sensing techniques, laser range finders, or stereo matching. The first successful SLAM algorithm using a monocular camera was called MonoSLAM [19]. The accuracy obtained using SLAM in a restricted area was encouraging. MonoSLAM resolved the depth problem by initializing the scale using objects with known size in the scene. The concept using objects of known size has been further developed by [20], wherein a database of object classes with their measured sizes was used. While doing SLAM, the objects found in the images were matched with the most suitable object class and the size of the object assigned as the class' size. However, both aforementioned methods rely on a priori knowledge or preparation and are therefore quite restrictive for an indoor SLAM. A much-used solution is to parameterize the distance by its inverse [21]. The inverse depth method estimates the depth as one state variable in the extended Kalman filter (EKF) pose estimation process using both close and very distant features. The reason for being able to include both nearby and distant features into the computation is the explicit parametrization of the inverse depth of a feature along a semi-infinite ray from the first detection pose. Thereby, the depth uncertainty may be modeled as Gaussian and included in the EKF process.

(a)

(b)

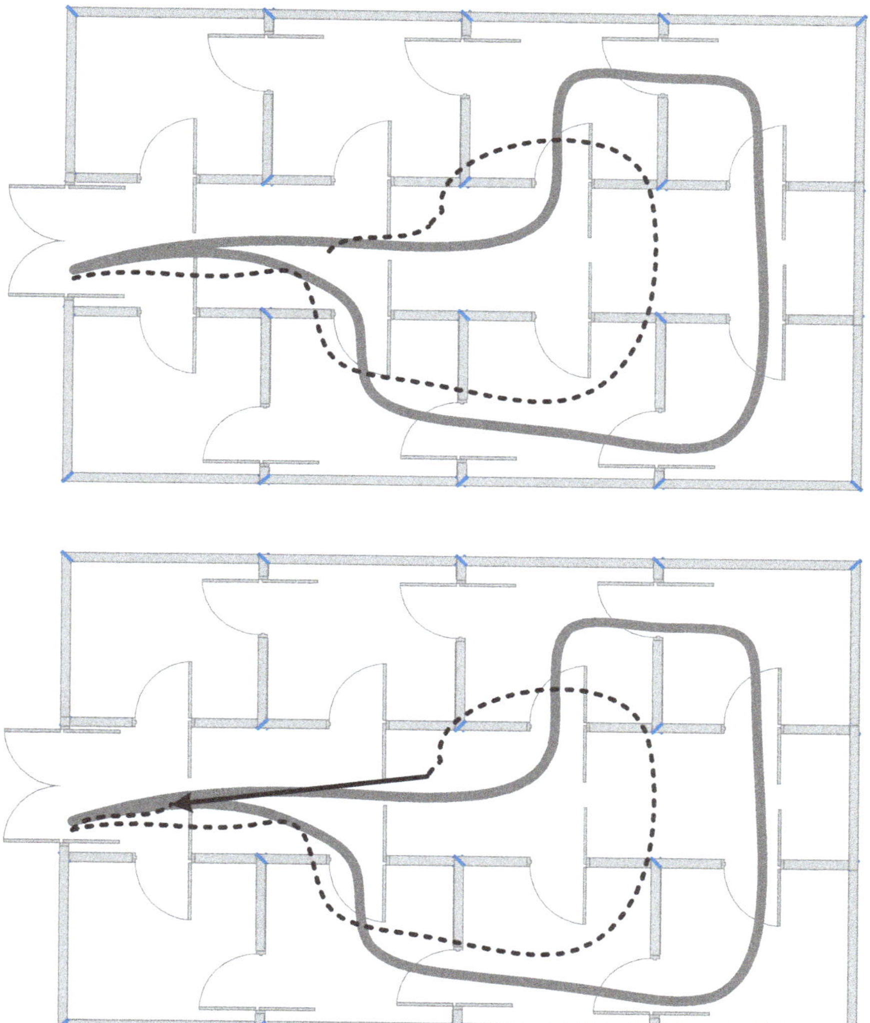

Figure 4.6 Loop closure in a SLAM system: (a) the solution obtained without loop closure (dotted line) drifts from the ground truth (gray solid line), and (b) the algorithm has detected a revisit and corrected the location (solid dark line).

At the time of writing, the ORB-SLAM method family was considered to be the state of the art of SLAM, and especially ORB-SLAM3 [22]. It is the first real-time SLAM and is able to compute a pure visual SLAM solution using monocular, stereo, and RGB-D cameras as well as a fused visul-inertial solution. As mentioned already and presented in Table 4.2, ORB-SLAMs are indirect methods, based on extracting ORB key-point features. Direct methods LSD-SLAM and DSO, as well in Table 4.2, are considered as the state of the art in their method classes.

Currently, deep learning visual SLAM techniques are used to reduce computation time and increase accuracy. State-of-the-art deep learning techniques can be used to improve SLAM processes throughout the whole pipeline, localization, mapping, and global optimization including loop closure [23]. Developing a full deep learning based SLAM pipeline is a challenging task; however, end-to-end solutions for the localization have already emerged, such as [24] for computing a visual localization solution and [25] for inertial and visual fusion. As the development of supervised deep learning models requires huge amounts of data, research has rapidly shifted toward developing self- [26] or unsupervised [27] deep neural networks for SLAM also utilizing reinforcement learning [28] due to its adaptability to a real dynamic environment. Interestingly, some of the state of-the-art methods still combine Kalman filters with end-to-end deep learning based localization, such as [29].

4.2.3 SLAM with Nonvisual Positioning Data

The motion of an imaging sensor can be inferred from successive image frames, which makes it possible to implement SLAM without other sensors. However, it is often beneficial to include an IMU or another type of motion sensor, such as wheel odometry in the case of robotics. Such externally measured motion information is usually incorporated to SLAM in the form of a control input.

Although SLAM is often associated with systems equipped by a camera or lidar, the concept can be applied to a variety of exteroceptive measurements that are not related to visual perception. Moreover, SLAM algorithms exist that only rely on loop-closure detection without mapping external landmarks. Some nonvisual SLAM approaches are discussed in the following sections.

4.2.3.1 Magnetic SLAM

In constructed environments, the abundance of local magnetic distortions makes the use of magnetometers very unreliable for heading determination. However, many of such distortions stay in the same place permanently, such as the effect of a steel bar in reinforced concrete. With appropriate instrumentation it is possible to track the magnetic anomalies with a SLAM approach.

For the mapping part of magnetic SLAM, a key problem is how to represent and estimate the ambient magnetic field: in essence, it is a vector field subject to Maxwell's equations and thus spatially correlated. A popular approach is to apply *Gaussian process* regression. A Gaussian process is defined as a stochastic process f with mean and covariance functions μ and $\kappa\left(x, x'\right) = \operatorname{cov}\left(f(x), f(x')\right)$, respectively [30]:

$$f(x) \sim GP\left(\mu(x), \kappa\left(x, x'\right)\right) \tag{4.7}$$

if the joint distributions of the values of f evaluated at N points is multivariate Gaussian for any finite N:

$$\begin{bmatrix} f(x_1) \\ \vdots \\ f(x_N) \end{bmatrix} \sim \mathcal{N}\left(\begin{bmatrix} \mu(x_1) \\ \vdots \\ \mu(x_N) \end{bmatrix}, \begin{bmatrix} \kappa(x_1, x_1) & \cdots & \kappa(x_1, x_N) \\ \vdots & \ddots & \vdots \\ \kappa(x_N, x_1) & \cdots & \kappa(x_N, x_N) \end{bmatrix} \right). \tag{4.8}$$

Similarly to regression with regular normal distributions, one can fit a Gaussian process onto (noisy) measurements. When modeling magnetic fields, a common choice for the covariance function κ are squared exponential covariance functions of the form

$$\kappa\left(x, x'\right) = \sigma^2 \exp\left(-\frac{\left\|x - x'\right\|^2}{l^2}\right), \tag{4.9}$$

with hyperparameters σ and l defining the standard deviation and length scale, respectively; then, the unknown magnetic map is obtained by estimating the mean function $\boldsymbol{\mu}(\boldsymbol{x}) = [m_E(\boldsymbol{x}) \quad m_N(\boldsymbol{x}) \quad m_U(\boldsymbol{x})]^T$ where $\boldsymbol{x}$ denotes the position, and m_E, m_N, m_U are the magnetic field strengths along the (map) coordinate axes (chosen as East–North–Up in this example). Thus, just like visual SLAM, magnetic SLAM requires estimating the orientation in addition to the position in order to resolve the magnetometer measurements in the map coordinate frame. In practice, it may be necessary to divide the area of operation into smaller cells and estimate a separate magnetic map for each cell to maintain scalability to larger environments [31].

Often, a magnetic SLAM system is implemented by combining a magnetometer with another sensor to obtain motion information; this can be, for example, PDR for pedestrian applications or wheel odometry in robotics. However, a sufficiently nonuniform magnetic field can be leveraged to derive motion information: using an array of magnetometers with known geometry, changes in the time series of magnetometer measurements due to user motion can be separated from those caused by changing local magnetic disturbances, enabling magnetic odometry [32].

4.2.3.2 Radio SLAM

Using radio signals as landmarks is an intuitive way to implement SLAM: beacons such as WLAN APs can be identified by the MAC address, they stay in the same location for a long time, and their signal can be observed in a certain coverage area, which makes them analogous to visual landmarks. The mapping can be implemented in terms of a fingerprint map or by estimating the coordinates of the individual transmitters.

The main motivation of taking the fingerprinting approach (i.e., creating a map of RSSI values) is to avoid the challenges of non-line-of-sight (NLOS) propagation and channel modeling, but at the cost of the amount of landmarks to be estimated. Similarly to magnetic SLAM (Section 4.2.3.1), Gaussian process regression can be applied to radio fingerprint mapping [33]; a key difference between radio and magnetic mapping is that the magnetic field vector always consists of three components while the user may observe an arbitrary number of radio signals. Using Gaussian processes is computationally heavy and does not scale well to large maps; alternative solutions include, for instance, applying a piecewise linear signal model [34], and using the GraphSLAM algorithm [35].

Estimating the transmitter coordinates keeps the number of landmarks much smaller than the fingerprinting approach and, in principle, basic algorithms such as EKF-SLAM [12] can be applied. In practice, this approach needs range measurements to the beacons. In the context of WLAN signals, attenuation models have traditionally been utilized to convert the RSSI to a distance, which has also been applied to SLAM [36]. However, TOA measurements, including RTT with modern WLAN infrastructure, offer more robust ranging as the measurement is not sensitive to changing attenuations caused by moving objects or the body of the user.

4.2.3.3 Trajectory Mapping by Loop Closure Only

The principle of SLAM has also been applied to the scenario where the user only has motion sensors (e.g., an IMU) but no capability of exteroceptive sensing to observe landmarks. In this case, the mapping is based on learning motion patterns when the user returns to a previously visited location.

The FootSLAM algorithm [37] creates the map as a dynamic Bayesian network where the probabilistic map consists of transition probabilities between uniform hexagonal cells covering the area of operation. The rationale behind the algorithm is that the probability of transition to one of the six neighboring hexagons is not uniform but subject to *physical constraints* such as walls and *visual cues* that affect the user's decision on the next step. Although the algorithm has no means to observe these factors, it can use the consistency of motion as a weighting criterion for the user position and probabilistic map whenever a cell is revisited: when implemented using a PF, the history of transitions from one map cell to another is stored for each particle. Then, if the ith particle crosses the eth edge ($e \in \{1, \ldots, 6\}$) of the cell C at time t, the particle's weight is updated [37]:

$$w_t^{(i)} \propto w_{t-1}^{(i)} \frac{N_{C,e}^{(i)} + \alpha_{C,e}}{\sum_{j=1}^{6} \left(N_{C,j}^{(i)} + \alpha_{C,j} \right)} \tag{4.10}$$

where $N_{C,j}^{(i)}$ denotes the total number of times the ith particle has crossed the jth edge of cell C (in either direction), and $\alpha_{C,j}$ is an a priori parameter.

As long as the trajectory contains a sufficient number of loop closures (i.e., revisits to a previously occupied map cell), FootSLAM can mitigate IMU error accumulation and maintain a stable navigation solution, assuming the measurement errors can be sufficiently well modeled [37]. Note that in this framework, there is no separate algorithm to detect loop closures; the update (4.10) is simply triggered whenever the position of a particle transitions from one cell to another.

4.3 Cooperative Navigation

Cooperative navigation (CN, also known as collaborative or peer-to-peer navigation) is an approach where the positioning accuracy of a navigation unit is improved by utilizing the position estimates, and possibly other information, determined by other units in the same area. CN is based on two central assumptions: first, that each individual user is able to independently estimate their navigation state and state uncertainty, and second, that they have the ability to measure or estimate range and range uncertainty to other cooperating users, preferably also the mutual direction, and communicate the result. This phase is called the *measurement phase*. In the following phase, the *location update phase*, an updated state is computed using the shared information. By exchanging real-time estimates of state, relative range, and state uncertainty, each cooperating user reduces the rate at which the error of the group of cooperating users accumulates, or even enables the calculation of the location estimate in conditions where it would not otherwise be possible. An example of the latter situation is sharing the GNSS positioning solution of users outdoors to their cooperators indoors. Examples of the ancillary data that might be shared are GNSS differential corrections or SBAS messages, and thereby all units do not require a specialized GNSS receiver.

Here, we will concentrate on the methods contributing to the computation of the cooperative navigation solution and will leave the aspect of controlling cooperative systems to other forums. Cooperative navigation gathers well together all topics discussed previously in this book; sensor measurements and computer vision are used for computing the position solution for individual users, radio positioning means for ranging between cooperating users, and finally computing the cooperative improved navigation solution for some or all users with estimation. However, the task is challenging because in cooperative navigation the measurements are the distances and angles between pairs of users with possibly erroneous locations, and each user's initial location is unknown for others.

4.3.1 Centralized and Noncentralized Calculation

Calculation of the actual cooperative location estimate in the *location update phase* can be implemented in two ways: centralized or noncentralized. The two setups are shown in Figure 4.7. In the centralized process, all users send the mutual distance measurements, and possibly additional data, to a central calculation unit. The

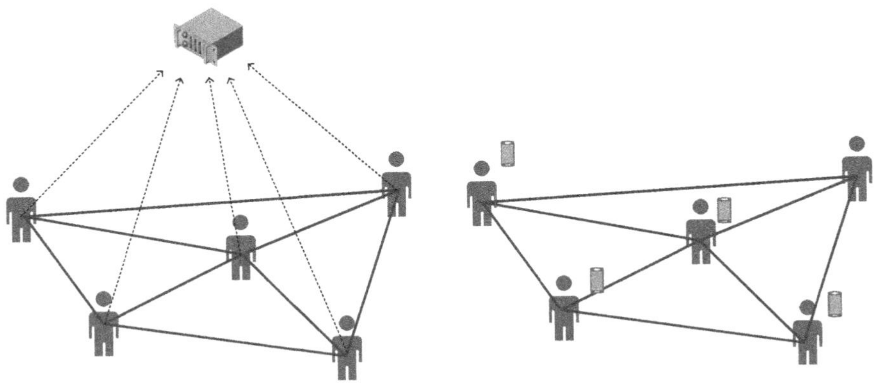

Figure 4.7 Decentralized and centralized setups for computing a cooperative location estimate.

central unit, for example a server, calculates the location estimates for each user and sends the information back to them. Noncentralized calculation is done in a distributed manner; each user calculates their estimates independently utilizing the location estimates of nearby users and the related distance measurements. Calculation methods may also be combinations of these two approaches.

In the centralized calculation, range measurements made by all units to other units are available. Such a setup enables precise location estimation for the whole group. However, abundant communication back and forth between the units and the central computing unit is demanding both in terms of power consumption and communication scheduling. Depending on the application area, it may be sufficient that the location estimates are not communicated back to all units, but only ones requiring the information. The drawback of a centralized calculation is that the process is relatively slow, and as the number of users increases, the positioning accuracy decreases due to communication delay. However, for smaller numbers of units, centralized cooperative navigation provides improved accuracy.

In the noncentralized calculation, each unit independently calculates the improved location estimate using the range measurements and communicates this estimate to the other units, if needed. Depending on the ranging method, the measurements might be available only to the nearby users and therefore the information is restricted compared to the centralized solution. Noncentralized algorithms can be divided into two categories: triangulation or iterative methods, such as Kalman filtering. The algorithms are often scalable, and thus the same computing method can be used for both large and small numbers of users [38]. Usually, when the communication between users is unreliable, the noncentralized solution is adopted as centralized processing of data with reasonable latency is not feasible.

As mentioned above, the calculation can also be implemented by combining the centralized and noncentralized methods. In such a setup, each user processes its own location estimate independently and then sends the complete solution to the central computing unit. When the central unit also receives the range measurements between the users, it is able to refine the location estimates. Users might be divided into smaller clusters, which first process their cluster's individual location estimation and then share the information with other nearby clusters [39].

4.3.2 Measuring the Range Between Users

The range between users is measured by transmitting radio signals, for example UWB or Wi-Fi and computing range using the RTT of the transmitted and received signal or any other range determination methods discussed in Chapter 3. According to [40], UWB can provide more accurate ranging than peer-to-peer Wi-Fi RTT, but the actual improvement depends on factors within the environment, such as building construction material and floor plans. Also, something to be considered in the system implementation is that at the time of writing, the allowed UWB signal power was regulated by some nations.

Here, we will give an example of computing the range using RTT and UWB signals used in a cooperative navigation. RTT is the duration in milliseconds (ms) it takes for a signal to travel from a starting point to a destination $(TT1 - TI1)$ and back to the starting point $(TI2 - TT2)$. The range (r) is measured by transmitting signals using UWB sensors and computing the RTT of the transmitted and received signal and multiplying that with the speed of light (c), as $r = c * RTT/2$. A simple RTT calculation method is

$$RTT = (TI2 - TI1) - (TT2 - TT1), \tag{4.11}$$

Usually the range is calculated with estimating RTT as an average over multiple transmissions to improve accuracy.

A more realistic scenario is one where the cooperation involves multiple users and the signals transmitted are used for providing other information in addition to ranging. In an example setup [41], a minimum of five users are required and the cooperative navigation solution is computed in a decentralized manner. As the example uses UWB for ranging, the line-of-sight signals are preferred for good performance. Therefore, all users periodically announce their presence to other users in the area and keep an updated list of others visible. It is possible to create a ranging and communication setup to up to 32 nodes through a method called time slicing based on a user node addresses. For time slicing, user nodes are synchronized to a common time base by using GNSS for initialization. By using a local oscillator, the time is stable for over 30 minutes indoors.

Figure 4.8 shows a RTT computation flow, where an individual user starts transmitting position information $(TI1)$ to announce its presence to other nodes, and receives a reply position from the receiver submitted at $(TT1)$. The ranging

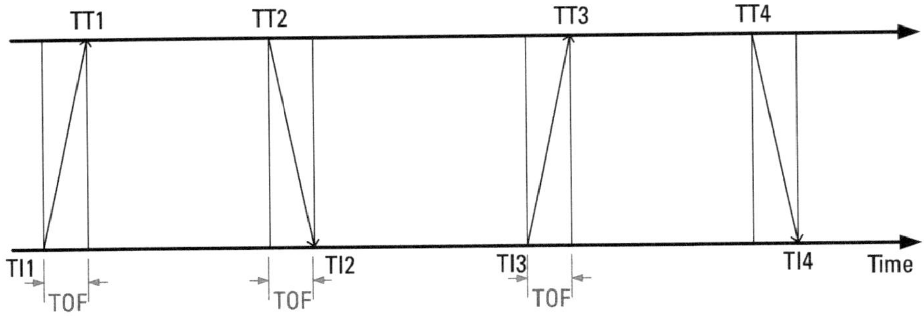

Figure 4.8 Computing the RTT solution.

cycle is divided into four subcycles in order to exchange sufficient timing data to allow the computation of the TOF measurement. The time used for the first two subcycles is computed as $\mathrm{T}_{RoundTripA} = TI2 - TI1$, and the delay between the second and third subcycles as $\mathrm{T}_{ReplyA} = TI3 - TI2$. Calculations are done similarly for the following two subcycles and denoted here with B. During the subcycles all users exchange information on their current position estimates and its uncertainty. During the process additional information about users not in the LOS range is shared to also provide such users situational awareness, namely knowledge about the cooperators' states. Additional information may also include state flags about the user, for example if the user is presently static (ZUPT condition) or has an independent ability to measure its position such as a GNSS fix. The RTT solution is computed as

$$\begin{aligned} \mathrm{T}_{RoundTripA} &= TI2 - TI1 \\ \mathrm{T}_{RoundTripB} &= TT3 - TT2 \\ \mathrm{T}_{ReplyA} &= TI3 - TI2 \\ \mathrm{T}_{ReplyB} &= TT2 - TT1 \\ \mathbf{RTT} = 4 \cdot TOF &= \mathrm{T}_{ReplyB}(\mathrm{T}_{RoundTripA} - \mathrm{T}_{ReplyA}) \\ &\quad + (\mathrm{T}_{RoundTripB} - \mathrm{T}_{ReplyB}) \end{aligned} \tag{4.12}$$

4.3.3 Computing the Cooperative Navigation Solution

In decentralized computation, the range information (RTT) from the user (u) to a cooperator (c) may be included into the navigation solution for example via the measurement model $\mathbf{z}_k = \mathbf{H}_k\mathbf{x}_k$ discussed in Chapter 3. In our example, the user is navigating indoors and measuring the horizontal motion using inertial sensors and vertical motion with a barometer. The cooperative user state $\mathbf{x}_k$ is defined as a 16-element vector including the user's position (x, y, z), velocity

(v_x, v_y, v_z), attitude $(pitch, roll, yaw)$, gyroscope bias (g_x, g_y, g_z), accelerometer bias (a_x, a_y, a_z), and barometer bias (b_z). Observation matrix $\mathbf{H}_k$ is

$$\mathbf{H}_k = [d_x d_p, d_y d_p, d_z d_p, 0, ..., 0]. \quad (4.13)$$

The $d_i d_p$ term is calculated from the predicted position coordinates $i = x, y, z$ of both the user (u) and the cooperator (c) as

$$d_x d_p = \frac{x_u - x_c}{R_p red}, \quad (4.14)$$

where $R_p red = \sqrt{(x_u - x_c)^2 + (y_u - y_c)^2 + (z_u - z_c)^2}$.

Since the measurement model makes direct use of the estimated position states of both the user and the cooperator, the formed vector is both scaled and rotated by position state errors in either user. The state's reliability estimate may be used in the process by ignoring observations from users whose state is not reliable to speed up and simplify the calculation. However, when the number of users is small, the use of even low-quality data can be useful to get a solution. The optimal approach is to adaptively select the data to be used based on the number of available cooperators and the reliability of their data.

Alternatively, the UWB range measurements may be used for obtaining the cooperative navigation solution by estimating location and heading offsets between the local coordinate frames of the user and the cooperator, assuming that the vertical axes are aligned [42]. In such a setting, the user's state vector may be presented as $[x_k y_k z_k \delta x_k \delta y_k \delta z_k \gamma]^T$, where (x_k, y_k, z_k) are the local coordinates, $\delta x_k \delta y_k \delta z_k$ the coordinate offsets between the user and cooperator in the local coordinates, and γ the heading offset between the local coordinate frame's of the user and cooperator. The range measurements are now entered into the model using the observation matrix (H_k) defined as

$$H_k = \begin{bmatrix} \sqrt{\delta x_k^2 + \delta y_k^2 + \delta z_k^2} \\ z_k \end{bmatrix} \quad (4.15)$$

where the first row of the matrix is the measured distance between the user and the cooperator and the second is the users estimated vertical position.

4.4 Computer Vision-Based Tracking

Tracking is a process of following something, or someone, dynamic and originally done by radars. The difference to navigation is that the position and motion information is not computed by the target, but by a third party. The radio localization methods discussed throughout this book are inherently tracking solutions. This means that the navigation solution is not usually computed in the user device, but remotely. However, solutions of such methods are still mainly meant for resolving the user's egomotion. Fast emerging radio tracking application is, for example, the coordination of autonomous robots in a warehouse. Here, we will discuss the methods and challenges in tracking using static cameras and computer vision methods.

Tracking is one of the most challenging computer vision tasks and relates to locating an object, mainly a human in an indoor navigation scene, over a sequence of images. The main related applications are in monitoring and surveillance, but also augmented reality and sports. Tracking research began in 1981 by Lucas and Kanade [43] and at present, the state-of-the-art methods are Siamese deep learning based trackers [44]. Today the most updated platform for the tracking research and development is the visual object tracking (VOT) challenges portal [45], which provides the visual tracking community benchmark datasets, methods and publications, as well as a forum for discussions and workshops. The definition of visual tracking, according to VOT, is that no prior knowledge about the object tracked is available. Here, we concentrate on 2D tracking, although 3D tracking is a relevant research topic as the development of autonomous systems is active. However, 3D tracking relates heavily to object depth estimation and therefore is not the focus of this section.

Tracking includes various steps of detecting an object to be tracked, understanding and following its motion, and relocating it after occlusions. Tracking can be done for single or multiple objects simultaneously. The existing tracking algorithms may be categorized in many ways, two of which are categorization based on the optimization of objects' trajectory (offline and online), or based on how they find the tracked object from the image into using templates (i.e., generative or discriminative, or collaborative) [46]. Offline tracking means that the captured images are scanned back and forth to improve the tracking solution, as the online tracking is done in real time and is therefore more interesting for the navigation domain. However, the methods for processing the data are largely the same. Generative trackers are divided into filter-based and template matching and are based on first detecting an object, representing it with a bounding box, and then searching for similar regions from consecutive images. Discriminative trackers use machine learning for separating the object from the background in the image, and collaborative models fuse algorithms from the two other categories.

4.4.1 Tracking Pipeline

The most important components of an object tracker are [47] (1) object detection, (2) object representation, (3) predicting position and motion of the tracked object, (4) matching similarity and optimization, and (5) model updating.

The way the detected object is defined varies: by one or multiple bounding boxes like in Figure 4.9, forming template-based methods, an area in the image,

Figure 4.9 Bounding box showing a detection of a human from an image. Numbers in the upper left corner of the box show its location in pixels.

or a fine-grained object. The main difficulty in tracking is the changes in the appearance of the object due to changes in the setup—rotation, scaling, illumination changes, motion blur, and especially occlusions—that possibly lead to the disappearance of the object from the images and a momentary total loss of tracking. Therefore, modern detectors are adaptive using predictions from previous image frames. The object's representation model's characteristics need to be optimized for it to be able to keep track of the object over consecutive images. If the model is too flexible, it will drift away from the correct presentation, and if it is too rigid, it will lose the object when its appearance changes. Traditional trackers used consist mainly of three types of data structures for representation: 2D matrices of brightness values, color histograms, or feature vectors. At present, tracking methods are increasingly based on deep learning, where the objects are detected and represented, or the full tracking pipeline, implemented with convolutional neural networks.

To keep the object in track, its motion and position need to be predicted from image sequence. Traditionally, mainly the Kalman filtering discussed in Section 3.3 has been used for the task due to its simplicity and fast processing. For each frame the results have been obtained for object detection and prediction of the motion and position of the object we have been tracking throughout previous images. These must be matched and the result optimized for a correct tracking solution. In the case of single object tracking, matching the objects across image sequence is quite straightforward. Matching can be done, for example, by using calculated motion and position information to restrict the present image's search area for an object found in the previous image.

When tracking multiple objects, additional information is needed to distinguish the objects and find their corresponding matches over the image sequence. Traditionally matching in multiobject tracking, also called assignment, has been done using the Hungarian algorithm. The Hungarian algorithm, also called bipartite graph matching, computes the optimal solution for matching the objects. In any bipartite graph, the number of edges in a maximum matching equals the number of vertices in a minimum vertex cover. To find this minimum, we define a cost matrix C of size $|S| \times |T|$ consisting of S the detected objects and T the predicted positions. Now, the edge weights in the cost matrix C present how much the positions of the detected objects in the present image and their estimated positions overlap. The algorithm builds a graph with the edges (E), and assigns vertices (nodes) (V) from one side of the graph to the other. The maximization may be implemented by using for example the *Dijkstra algorithm* with the time complexity $O(V^2 \log V + VE)$. The last component of the tracking algorithm is updating the model to deal with the variation in the appearance. Various methods for the update exist; traditionally, for example, principal component analysis (PCA) has been used.

4.4.2 The Future of Tracking

Deep learning has also started to dominate the tracking domain [48]. Deep learning based object detection is by far the best way of finding a tracked object from images as they are more invariant for appearance variations than traditional algorithms. Also, prediction of the motion of the objects across image sequence is done robustly with deep learning, especially in the case of tracking multiple objects and suffering from overlapping objects and occlusions. Simple online and realtime tracking with a deep association metric (DeepSort) [49] is a CNN architecture for multiobject tracking. It joins the object detection and appearance presentation obtained using deep learning with Kalman filtering and the Hungarian algorithm in a deep neural network. However, when the objects have a similar appearance and complex patterns of the movement the basic DeepSort fails to differentiate the identity of objects. To address these issues, the state-of-the-art multiobject tracking algorithms are based on Siamese networks, which are artificial neural networks learning a similarity function. They simultaneously process two different input vectors to compute comparable outputs via using shared weights in the network [50]. Despite all discussed solutions, tracking is still the most challenging computer vision task and one that is far from being perfected. Therefore, the scientific community is looking to new innovations based on the most recent leap in deep learning architecture design: visual transformers. The first transformer-based trackers [51] have started to emerge at the time of writing.

4.5 Radio-Based Indoor Positioning

In the following, we approach the description of a radio-based indoor positioning system through an extensive simulated system demonstration. We consider aspects of obtaining positioning-related measurements such as AOA/AOD and TOA, and present a fundamental EKF-based filtering solution to estimate and track a moving user device. In addition, we discuss the most relevant aspects of channel modeling required for simulating radio-based indoor positioning systems.

Compared to positioning systems utilized in an open-space environment, indoor positioning systems are often operating in compact spaces with unique interior features and frequently compromised availability of the important LOS signal. Moreover, the performance of a radio-based positioning system is usually subject to a specific floor plan as well as on the construction materials used for the walls, floors, and ceilings. For positioning it is possible to utilize traditional positioning-related measurements, such as AOA, AOD, TOA, TDOA, and RTT. However, in the most traditional positioning setting, the abovementioned positioning-related measurements are dependent on the LOS signal availability, which provides correct geometric relations between the reference APs and the user device, and thus enables accurate position estimation. Nonetheless, at the cost of substantially increased computational complexity it is also possible to utilize NLOS signals for positioning, as described, for example, in [52].

4.5.1 Channel Modeling

Due to historical reasons, many of the channel models presented in the literature, for example [53–55], are designed for outdoor usage and often oversimplify the radio propagation characteristics of indoor scenarios. In addition, several channel models targeted at performance evaluations of communication systems, such as [56, 57], exploit stochastic processes. Consequently, there is no true linkage to underlying channel geometry, including angles and ranges of radio paths as well as device velocities. Nonetheless, the majority of radio-based positioning methods are dependent on the channel geometry, and thus more elaborate radio channel models are needed. It should be emphasized that the channel itself is the key for radio-based positioning, as it conceals all the essential parameters needed for the positioning solution. That is, for each user device location, there is a specific channel outcome from which the user location can be estimated, either unambiguously or ambiguously.

Considering the complexity of indoor environments, with specific floor plans, furniture, and construction materials, it is extremely difficult to find well-generalized channel models for global employment, and therefore site-specific channel models are often preferred. To this end, ray-tracing-based channel models have recently gained much attention, for example, as part of 5G standardization [58]. Due to inherent geometric correctness of ray-tracing models with respect to pathwise angles and ranges for a specific indoor scenario, ray-tracing is extremely suitable for channel modeling in positioning-related studies. For interested readers, detailed information on ray-tracing models in the context of radio channel modeling can be found, for example, in [59].

From appropriate ray-tracing simulations, one can obtain the necessary channel parameters for generating signal-level channel responses. These parameters include, at least:

- Pathwise angular information as AOA and AOD, including the azimuth and elevation angles of each path;
- Pathwise propagation delays, which provide information on TOA and radio propagation distance for each path;
- Pathwise received powers and phase shifts, which depend on the radio propagation distance and possible interactions in the radio channel, such as reflection, diffraction, and/or scattering in the radio propagation path.

It is worth noting that the received power and phase shift can be also presented with a single complex valued channel coefficient by considering the polar form representation, where the square root of power is the absolute value (or modulus, or amplitude), and the phase is the argument of the complex number (i.e., $\sqrt{P}e^{i\varphi}$ where P is the power and φ is the phase).

Building on the ray-tracing-like channel description, where the channel is considered a composition of a set of observable radio propagation paths, it is possible to evaluate a channel impulse response. Assuming a transmit side antenna array with N_{TX} elements, and a receive side antenna array with N_{RX} elements, the antenna-element-wise channel impulse response $\mathrm{H}_{\mathrm{r}}(\tau) \in \mathbb{C}^{N_{\mathrm{RX}} \times N_{\mathrm{TX}}}$ as a function of propagation delay τ can be written as [58, 60]

$$\mathrm{H}_{\mathrm{t}}(\tau) = \sum_{k=0}^{K-1} h_k a_{\mathrm{RX}}(\eta_{\mathrm{AOA},k}) a_{\mathrm{TX}}^{H}(\eta_{\mathrm{AOD},k}) \delta(\tau - \tau_k) \tag{4.16}$$

where $(\cdot)^H$ denotes the Hermitian transpose, K is the number of multipaths, and h_k and τ_k are the complex path coefficient (i.e., amplitude/power and phase) and the propagation delay (i.e., TOA), respectively. Moreover, $\boldsymbol{\eta}_{\mathrm{AOA},k} = [\varphi_{\mathrm{AOA},k}, \theta_{\mathrm{AOA},k}]^T$, and $\boldsymbol{\eta}_{\mathrm{AOD},k} = [\varphi_{\mathrm{AOD},k}, \theta_{\mathrm{AOD},k}]^T$ are the AOA and AOD, consisting of azimuth angles $\varphi_{\mathrm{AOA},k}$ and $\varphi_{\mathrm{AOD},k}$, and elevation angles $\theta_{\mathrm{AOA},k}$ and $\theta_{\mathrm{AOD},k}$, for the kth path, respectively. The division of channel paths in time is handled by the Dirac delta function (i.e., a unit impulse), denoted as $\delta(\cdot)$. Furthermore, $\boldsymbol{a}_{\mathrm{TX}}(\cdot) \in \mathbb{C}^{N_{\mathrm{TX}}}$ and $\boldsymbol{a}_{\mathrm{RX}}(\cdot) \in \mathbb{C}^{N_{\mathrm{RX}}}$ are the antenna array specific steering vectors, which define the AOD and AOA dependent phases per antenna element with respect to the array center. Different arrangements of antenna array elements result in generally different steering vector realizations.

In addition to representing the radio channel in the time domain through an impulse response with time-multiplexed propagation paths, as given in (4.16), it is also possible to construct the radio channel in frequency domain. Considering again transmit and receive arrays of size N_{TX} and N_{RX}, respectively, the antenna-element-wise frequency response, denoted as $\mathrm{H_f}(f) \in \mathbb{C}^{N_{\mathrm{RX}} \times N_{\mathrm{TX}}}$ where f is the frequency, can be written as [52, 61]

$$\mathrm{H_f}(f) = \sum_{k=0}^{K-1} h_k e^{-j2\pi \tau_k f} \boldsymbol{a}_{\mathrm{RX}}(\eta_{\mathrm{AOA},k}) \boldsymbol{a}_{\mathrm{TX}}^H(\eta_{\mathrm{AOD},k}). \tag{4.17}$$

The above frequency response can be straightforwardly converted into discrete representation, for example, to use with OFDM-based signals. For this purpose, the channel frequency response can be further modified to represent a subcarrier-wise frequency response $\mathrm{H_{OFDM}}(n) \in \mathbb{C}^{N_{\mathrm{RX}} \times N_{\mathrm{TX}}}$ used for an OFDM-based transmission as

$$\mathrm{H_{OFDM}}(n) = \sum_{k=0}^{K-1} h_k e^{-\frac{j2\pi n \tau_k F_s}{N}} \boldsymbol{a}_{\mathrm{RX}}(\eta_{\mathrm{AOA},k}) \boldsymbol{a}_{\mathrm{TX}}^H(\eta_{\mathrm{AOD},k}), \tag{4.18}$$

where n is the subcarrier index, F_s is the sampling frequency, and N is the size of used fast Fourier transform (FFT) in the OFDM modulation.

The above channel representations, given in (4.16), (4.17), and (4.18), are formulated per a pair of transmit and receive antenna for antenna arrays of specific size. If the transmitter and receiver are implemented with parallel digital transmit and receiver RF chains, the above channel representations are usable as is. This type of MIMO processing at the transmitter and/or receiver is referred to as digital beamforming. However, in practice, transmitters and/or receivers are not always able to digitally process signals per antenna element, but over a group of elements after specific analog RF-stage preprocessing. In such a case, the transmitter and/or receiver is considered to apply analog beamforming. Moreover, if more than one digital sample is obtained, it is considered hybrid beamforming. The conventional analog RF-stage preprocessing considered with analog and hybrid beamformers involves antenna-element-wise phase manipulation similar to phased arrays. For example, assuming fully analog beamformers at the receiver and transmitter, the effective scalar channel impulse response for the MIMO channel in (4.16) is given as

$$h_t(\tau) = b_{RX}^H H_t(\tau) b_{TX}, \tag{4.19}$$

where $b_{TX} \in \mathbb{C}^{N_{TX}}$ and $b_{RX} \in \mathbb{C}^{N_{RX}}$ are the transmit and receive beamformers, respectively. When it is desired to target a beam in a specific direction, the beamformers can be determined according to steering vectors $a_{TX}(\cdot)$ and $a_{RX}(\cdot)$.

4.5.2 Description of the Simulated Positioning System

All simulations are performed by considering ray-tracing-based channel modeling in an indoor environment as shown in Figure 4.10. In the illustrated L-shaped room of 1,000 m^2 in area, there are three APs from which at least two are always visible when moving around the room. The exact (x, y, z)-coordinates of the APs are given as $(-39, -19, 4.9)$, $(-10, 19, 4.9)$, and $(-1, -19, 4.9)$. The user track considered for testing the positioning performance is represented by a smooth trajectory in the XY-plane with a fixed altitude (z-coordinate) of 1.5m. The user moves with a constant speed of 1 m/s while obtaining positioning measurements at intervals of $\Delta t = 0.25$s. The overall time duration of the track is 127.5s, which results in a total track length of 127.5m. Due to visibility reasons the ceiling is not shown in the floor plan, but it is included in the simulations while assuming a floor height of 5m.

For numerical result evaluations, we consider an OFDM-based positioning system with two fundamental system setups: one setup for the FR1, and another setup for the FR2. The main parameters used for the two simulation setups is shown in Table 4.3.

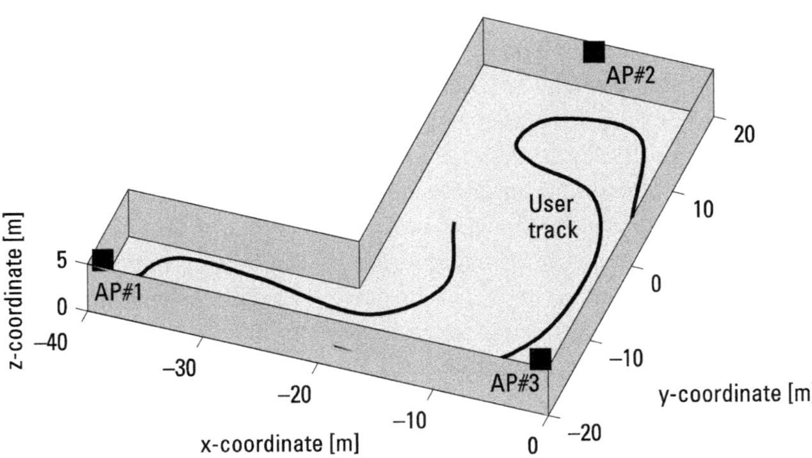

Figure 4.10 Indoor environment for the radio positioning demonstration.

Table 4.3
Main Parameters for Example Simulations

Parameter	FR1 setup	FR2 setup
Carrier frequency [GHz]	3	30
Bandwidth [MHz]	40	400
Subcarrier spacing [kHz]	30	120
FFT size	2048	4096
Number of active subcarriers	1200	3240
Number of antenna elements at AP	8	32
Number of antenna elements at user device	1	1
AP beamformer type	Digital	Analog
Antenna element types	Omnidirectional	Omnidirectional

Besides the two main setups, we show a few exceptional parameter combinations and vary different channel-related parameters to highlight specific aspects needed to take into account in the design and implementation of indoor positioning systems.

At each position, we employ a proprietary ray-tracing tool to obtain information on the current radio propagation paths, including pathwise angles, delays, phases, and powers. Based on the ray-tracing tool output, we exploit (4.18) to generate the deterministic ray-tracing-based channel response H_{OFDM}, and subsequently model the received signal vector at the nth subcarrier as

$$y_{\mathrm{RX}}(n) = \mathrm{H}_{\mathrm{OFDM}}(n)s_{\mathrm{TX}} + n_{\mathrm{RX}}(n) \in \mathbb{R}^{N_{\mathrm{RX}}}, \tag{4.20}$$

where each element of $y_{\mathrm{RX}}(n)$ determines the received signal for one antenna element. Moreover, $s_{\mathrm{TX}}(n)$ is the transmitted signal at the nth subcarrier, and $n_{\mathrm{RX}}(n) \sim \mathcal{N}(0, \Sigma_{\mathrm{RX}})$ denotes measurement noise. For simplicity, we consider traditional thermal noise, whose power in dBm within the signal band can be defined as [62]

$$P_{\mathrm{n}} = -174 + 10\log_{10}(B) + F_{\mathrm{dB}} \text{ [dBm]}, \tag{4.21}$$

where the value -174 is the approximate noise power spectral density at the room temperature, and B is the signal bandwidth. An illustration of ray-tracing-based propagation paths between one of the APs and a specific user position is shown in Figure 4.11.

4.5.3 Brief Description of the Measurements and the Utilized EKF

In order to take into account the user mobility, we utilize the EKF described in Chapter 3. Accordingly, we define a state-space model, consisting of a linear

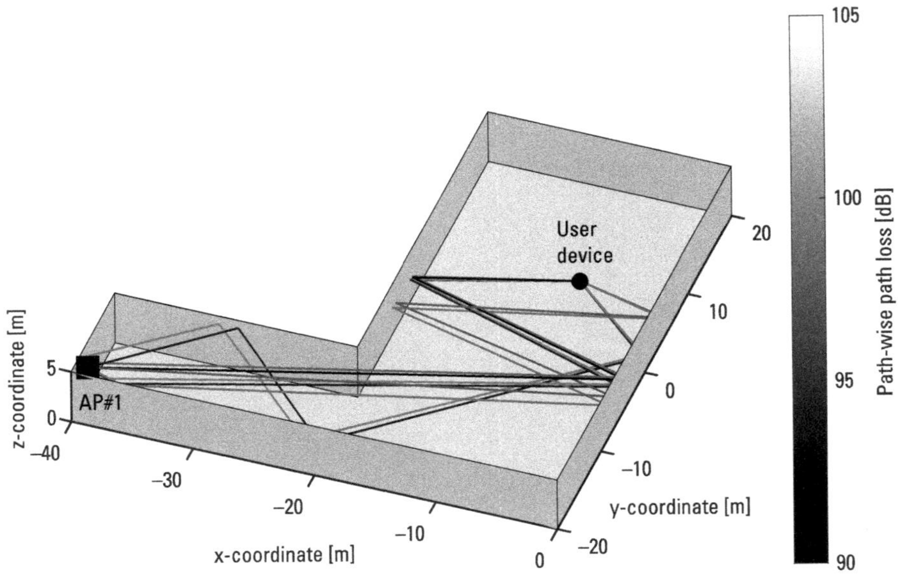

Figure 4.11 Illustration of ray-tracing-based propagation paths between AP and a specific user position.

state-transition model and nonlinear measurement model, as

$$s[k] = \mathrm{F}s[k-1] + w[k] \tag{4.22}$$

$$y[k] = h(s[k]) + n[k], \tag{4.23}$$

where $s[k]$ is the user state at time instant k, $\mathrm{F} \in \mathbb{R}^{6\times 6}$ is the state-transition matrix, and $w[k] \sim \mathcal{N}(0, \mathrm{Q}) \in \mathbb{R}^6$ is the process driving noise. In addition, $y[k]$ denotes the measurement vector, including, for example, TOA and/or AOA estimates from different APs at kth time instant. The measurement vector is affected by $h(\cdot)$, which describes the nonlinear measurement function as a function of user state, and $n \sim \mathcal{N}(0, \Sigma)$, which denotes measurement noise. Regarding the user state, we consider a simple state vector, given as

$$s[k] = \left[p_{\mathrm{u}}[k]^T \quad v_{\mathrm{u}}[k]^T\right]^T \in \mathbb{R}^6 \tag{4.24}$$

where $p_{\mathrm{u}}[k] = [x_{\mathrm{u}}[k], y_{\mathrm{u}}[k], z_{\mathrm{u}}[k]]^T$ is the 3D user position, and $v_{\mathrm{u}}[k] = [v_x[k], v_y[k], v_z[k]]^T$ is the 3D user velocity, defined according to directions of the x-axis, y-axis, and z-axis, respectively. Furthermore, the state-transition matrix is determined as

$$\mathrm{F} = \begin{bmatrix} \mathrm{I}_{3\times 3} & \Delta t \mathrm{I}_{3\times 3} \\ 0_{3\times 3} & \mathrm{I}_{3\times 3} \end{bmatrix}, \tag{4.25}$$

where Δt is the time duration between two consecutive time steps. We assume that the user moves with near constant velocity over a small time period, and consequently, utilizes a linear constant white-noise acceleration model, described in [63]. As a result, the process covariance matrix can be written as

$$Q = \begin{bmatrix} \frac{\Delta t^3}{3} I_{3\times 3} & \frac{\Delta t^2}{2} I_{3\times 3} \\ \frac{\Delta t^2}{2} I_{3\times 3} & \Delta t I_{3\times 3} \end{bmatrix} \in \mathbb{R}^{6\times 6}, \tag{4.26}$$

where σ_v^2 is the variance of user device acceleration for each velocity component axis.

As described in Chapter 3, the EKF performs recursive estimation by means of a prediction step and an update step. Regarding the considered linear state-transition model, the prediction step is straightforward. However, the update step, involving the nonlinear measurement function, is dependent on the type of available measurements and requires measurement-specific Jacobian matrices in order to obtain the required linear approximation of the measurement function. For the demonstrated system we assume obtaining TOA and AOA measurements at the AP side. In practice, TOA measurements can be obtained either by the user device or the AP, for example, through a RTT procedure. Moreover, depending on the beamformer type of the AP antenna array, the AOA measurements can be obtained, for example, using a multiple signal classification (MUSIC) algorithm or a beam-sweeping procedure. Nonetheless, assuming the use of TOA and AOA measurements, the measurement model for the mth AP at the kth time instant can be written as

$$\boldsymbol{h}(\boldsymbol{s}[k]) = \begin{bmatrix} \frac{\|\boldsymbol{p}_{\mathrm{u}}[k] - \boldsymbol{p}_{\mathrm{AN},m}\|}{c} \\ \mathrm{atan}_2\left(\Delta y_m[k], \Delta x_m[k]\right) \\ \arcsin\left(\Delta z_m[k]/d_m[k]\right) \end{bmatrix}, \tag{4.27}$$

where the first element refers to TOA, the second element to azimuth AOA, and the third element to elevation AOA. Moreover, $\Delta x_m[k] = x_{\mathrm{u}}[k] - x_m$, $\Delta y_m[k] = y_{\mathrm{u}}[k] - y_m$, and $\Delta z_m[k] = z_{\mathrm{u}}[k] - z_m$, where (x_m, y_m, z_m) is the position of the mth AP, and $\hat{d}_m[i] = \sqrt{\Delta x_m^2[k] + \Delta y_m^2[k] + \Delta z_m^2[k]}$ denotes the distance between the mth AP and the user device at time instant k.

Finally, assuming measurements from $m = 1, \ldots, M$ APs, the Jacobian matrix for taking the EKF update step can be written as

$$\mathrm{H}[k] = \begin{bmatrix} \mathrm{H}_{\mathrm{TOA}}[k] \\ \mathrm{H}_{\text{AZ-AOA}}[k] \\ \mathrm{H}_{\text{EL-AOA}}[k] \end{bmatrix}, \tag{4.28}$$

where the submatrices are further described as

$$\mathrm{H}_{\mathrm{TOA}}[k] = \begin{bmatrix} \frac{\Delta x_1[k]}{cd_1[k]} & \frac{\Delta y_1[k]}{cd_1[k]} & \frac{\Delta z_1[k]}{cd_1[k]} & \mathit{0}_{1\times 3} \\ \vdots & \vdots & \vdots & \vdots \\ \frac{\Delta x_M[k]}{cd_M[k]} & \frac{\Delta y_M[k]}{cd_M[k]} & \frac{\Delta z_M[k]}{cd_M[k]} & \mathit{0}_{1\times 3} \end{bmatrix}, \tag{4.29}$$

$$\mathrm{H}_{\text{AZ-AOA}}[k] = \begin{bmatrix} -\frac{\Delta y_1[k]}{d_{2\mathrm{D},1}[k]} & \frac{\Delta x_1[k]}{d_{2\mathrm{D},1}[k]} & 0 & \mathit{0}_{1\times 3} \\ \vdots & \vdots & \vdots & \vdots \\ -\frac{\Delta y_M[k]}{d_{2\mathrm{D},M}[k]} & \frac{\Delta x_M[k]}{d_{2\mathrm{D},M}[k]} & 0 & \mathit{0}_{1\times 3}, \end{bmatrix}, \text{ and} \tag{4.30}$$

$$\mathrm{H}_{\text{EL-AOA}}[k] = \begin{bmatrix} -\frac{\Delta x_1[k]\Delta z_1[k]}{d_1^2[i]d_{2\mathrm{D},1}[k]} & -\frac{\Delta y_1[k]\Delta z_1[k]}{d_1^2[k]d_{2\mathrm{D},1}[k]} & \frac{d_{2\mathrm{D},1}[k]}{d_1^2[k]} & \mathit{0}_{1\times 3} \\ \vdots & \vdots & \vdots & \vdots \\ -\frac{\Delta x_M[k]\Delta z_M[k]}{d_M^2[k]d_{2\mathrm{D},M}[k]} & -\frac{\Delta y_M[k]\Delta z_M[k]}{d_M^2[k]d_{2\mathrm{D,M}}[k]} & \frac{d_{2\mathrm{D},M}[k]}{d_M^2[k]} & \mathit{0}_{1\times 3} \end{bmatrix}. \tag{4.31}$$

In the above, $d_{2\mathrm{D},m}[k] = \sqrt{\Delta x_m^2[k] + \Delta y_m^2[k]}$ denotes the 2D (XY-plane) distance between the mth AP and the user device at time instant k. During the EKF iterations, Jacobian matrices are always evaluated with respect to the predicted state vector $s[k]$.

4.5.4 Positioning with CRB-Based Measurements

Utilization of theoretical CRB-based channel measurements, such as TOA and AOA, enable evaluation of positioning performance without the need for practical channel estimators. In addition, sampling TOA and AOA from known and well-defined distributions, such as Gaussian distribution, provides stability for the used positioning filter.

In the following results, we obtain samples of TOA and AOA by assuming Gaussian distributed measurement error with variances determined by the corresponding CRBs presented in Chapter 3. Besides parameters such as bandwidth and antenna array size, the CRB is affected by the SNR, which varies per AP along the user track. Furthermore, measurements are obtained only for the APs, which are in LOS with the user device.

In Figure 4.12, cumulative distributions of the range and AOA estimation error are shown for the CRB-based measurements considering the FR1 and FR2 parameter setups. It can be seen that the theoretical bounds are very low for both parameter setups. However, it should be remembered that the theoretical bounds are derived for ideal measurement conditions without any clock errors and for different transmitter or receiver impairments. In addition, only the

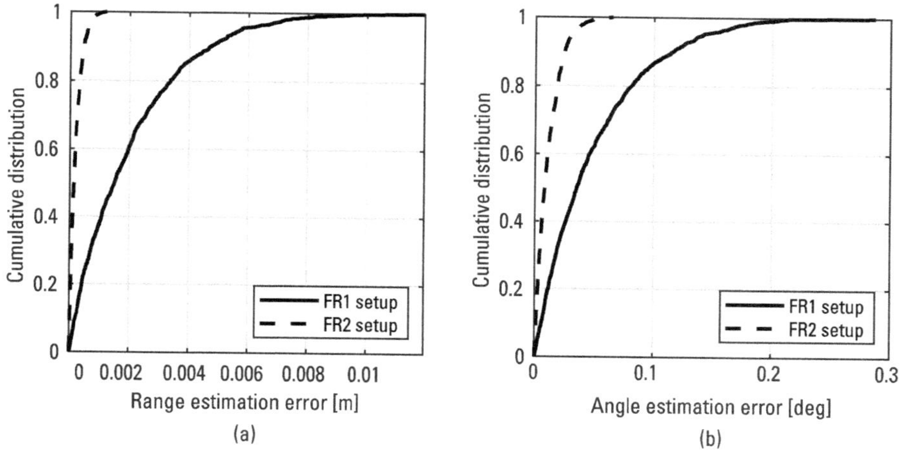

Figure 4.12 Cumulative distribution of (a) range and (b) AOA estimation error for CRB-based measurements.

LOS path is assumed in the channel, which is exceedingly optimistic for indoor scenarios.

Figure 4.13 shows the estimated user track using the EKF with the CRB-based measurements. Here, we consider only the FR1 setup, but obviously, similar results can also be obtained for the FR2. Due to the extremely high ranging and

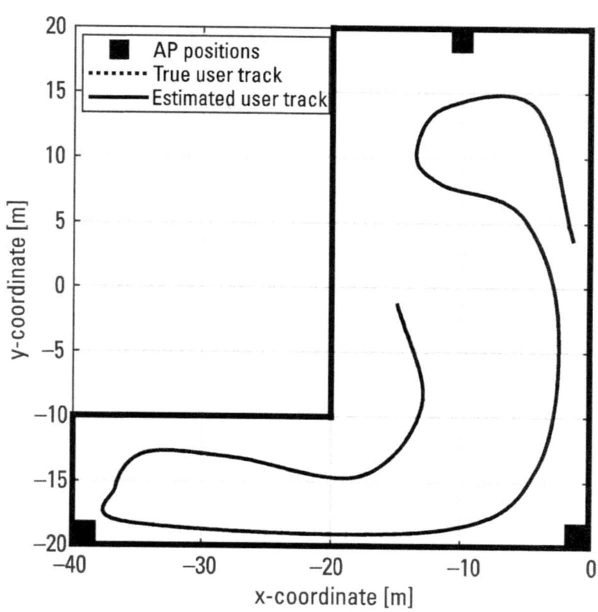

Figure 4.13 Illustration of the estimated user track using the EKF with the CRB-based measurements for FR1. Due to high positioning accuracy, the curve for the true user track is found below the estimated track curve.

angles estimation accuracy, the estimated user track lies practically on top of the true user track. Therefore, we further visualize the resulted positioning error through coordinate-wise position estimation accuracy, as shown in Figure 4.14. Whereas the x-coordinate and y-coordinate have a similar scale of error, the z-coordinate is subject to larger errors. This is due to the bad AP geometry regarding the elevation (z-axis), as all APs are located at the same height.

Similar to the evaluation of Fisher information and the related CRB for given measurements, it is also possible to define posterior CRB [64], which provides a bound for positioning accuracy during a filtering process, such as the EKF. This approach combines a priori information from the prediction step with the measurement-related information to obtain a posterior bound for each time step. However, for the sake of clarity of the representation, we do not consider posterior bounds further, but an interested reader can take a closer look in [64].

4.5.5 Positioning with Practical Channel Estimators

It is obvious that the results obtained with the CRB-based measurements are ideal and generally out of reach in practical scenarios. In this section we consider practical channel estimators for obtaining TOA and AOA measurements. We see that because of different nonideal conditions, for example, related to multipath propagation and detection of LOS conditions, the positioning task becomes much more challenging than the one shown with ideal CRB-based measurements.

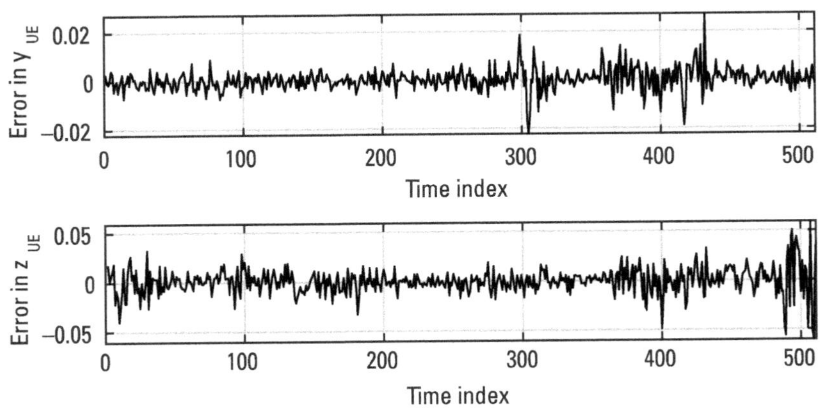

Figure 4.14 Illustration of the coordinate-wise position estimation accuracy.

First, we study the TOA and AOA estimation accuracy for practical estimators. Throughout the results, we estimate the TOA based on the frequency domain correlation [52, 65], which does not limit the estimation accuracy into to the sample time interval, as a conventional time correlation based estimator. Furthermore, regarding the FR1 setup with digital beamforming capability, the AOA is estimated using a conventional MUSIC algorithm, which has been further developed in the literature for improved performance [66–68]. Then, for the AOA estimation in the FR2 setup with analog beamformers, we employ a beam-sweeping based method considering beam power measurements [69, 70].

Figure 4.15 shows the the range estimation error for the FR1 and FR2 setups. Contrary to the corresponding accuracies with the CRB-based measurements, only the FR2 setup is able to provide consistent submeter ranging accuracy. The corresponding AOA estimation accuracies are shown in Figure 4.16, where it can be seen that the difference between the performance of the FR1 setup and FR2 setup has decreased. In order to highlight the effect of multipath interference on the estimation accuracy, we also evaluate the performance of the practical estimators by considering only the LOS path from the ray-tracing data. Moreover, the estimation accuracies for the range and AOA for illustrating the effect of multipath interference are shown in Figures 4.17 and 4.18, respectively.

It is clear that multipath propagation can considerably affect the estimation performance. An important observation in Figures 4.17 and 4.18 is the error floor visible for the results considering all multipaths. This error floor is due to fact that the estimators can tolerate multipaths within reason as long as the time and/or spatial characteristics of the multipaths differ from the desired LOS path. Otherwise, the estimator might misinterpret multiple paths as one.

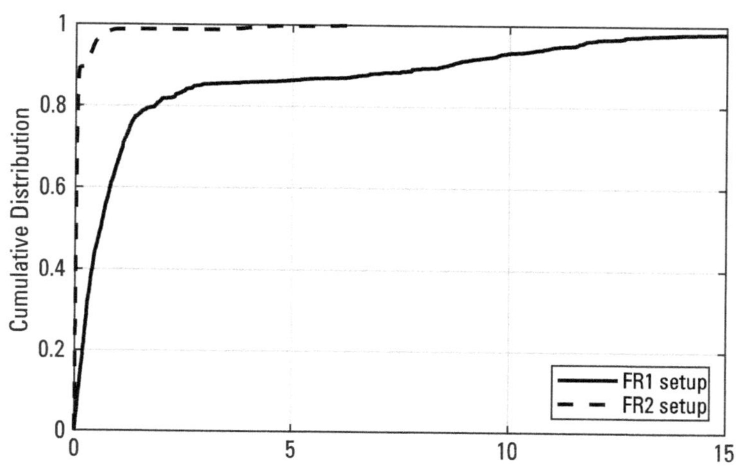

Figure 4.15 Cumulative distribution of range estimation error for practical channel measurements.

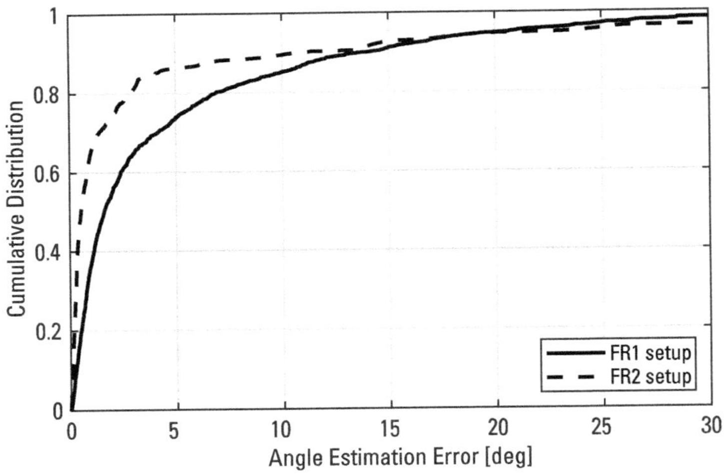

Figure 4.16 Cumulative distribution of AOA estimation error for practical channel measurements.

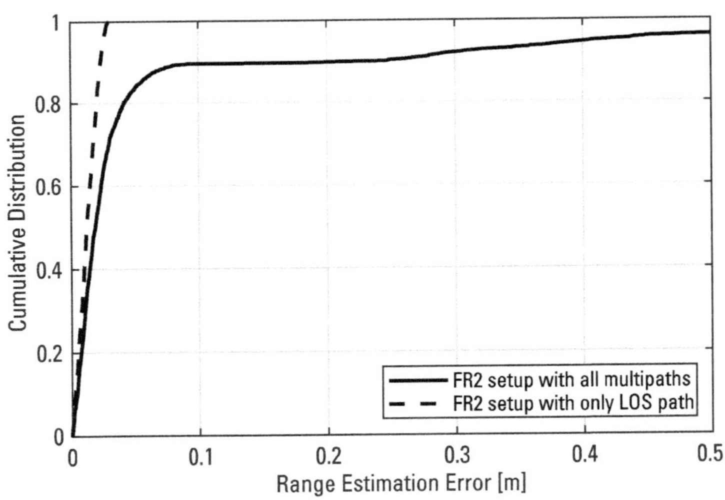

Figure 4.17 Cumulative distribution of range estimation error for illustrating the effect of multipath interference.

For obtaining the positioning results, we utilize a simple outlier detection scheme presented in [71]. The outlier detection is needed to first remove significant measurement outliers (e.g., due to multipath), and second, to manage detection of non-LOS occasions for discarding such measurements. In a case where the outlier detection is omitted, the positioning performance is radically dropped and there is a high probability that the EKF diverges.

Nonetheless, the estimated user tracks for the FR1 and FR2 setups are illustrated in Figures 4.19 and 4.20, respectively. As expected, based on the above

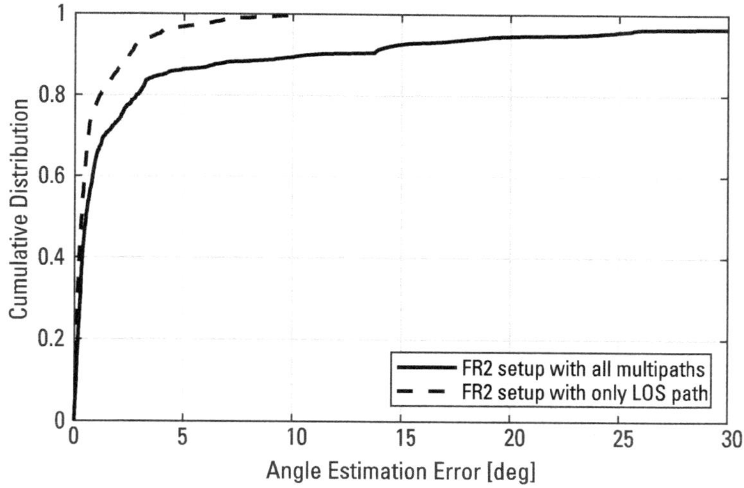

Figure 4.18 Cumulative distribution of angle estimation error for illustrating the effect of multipath interference.

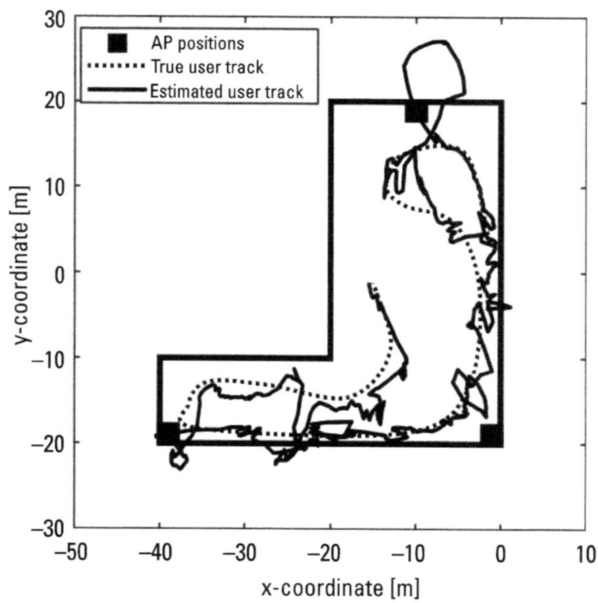

Figure 4.19 Illustration of the estimated user track using the EKF for FR1.

range and AOA estimation accuracies, the FR2 setup is able to provide a better performance. Furthermore, in order to visualize the 3D aspect of the positioning system, the estimated user track regarding the FR2 setup in Figure 4.20 is reillustrated in 3D space in Figure 4.21. The figure reveals the relatively poor estimation performance considering the *z*-coordinate. As indicated earlier with

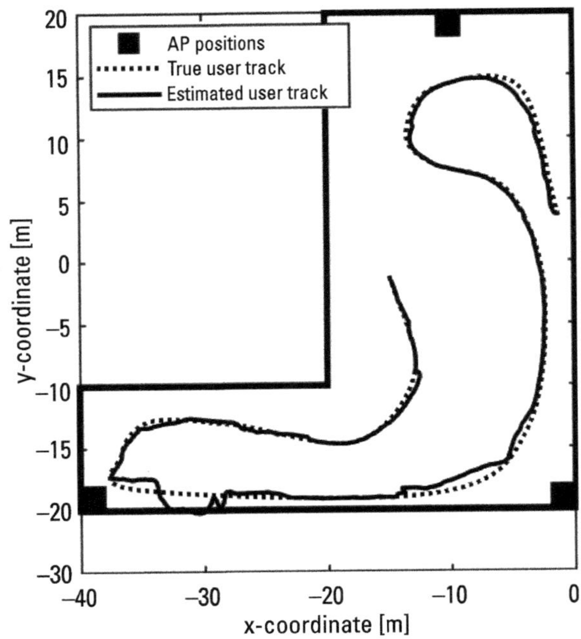

Figure 4.20 Illustration of the estimated user track using the EKF for FR2.

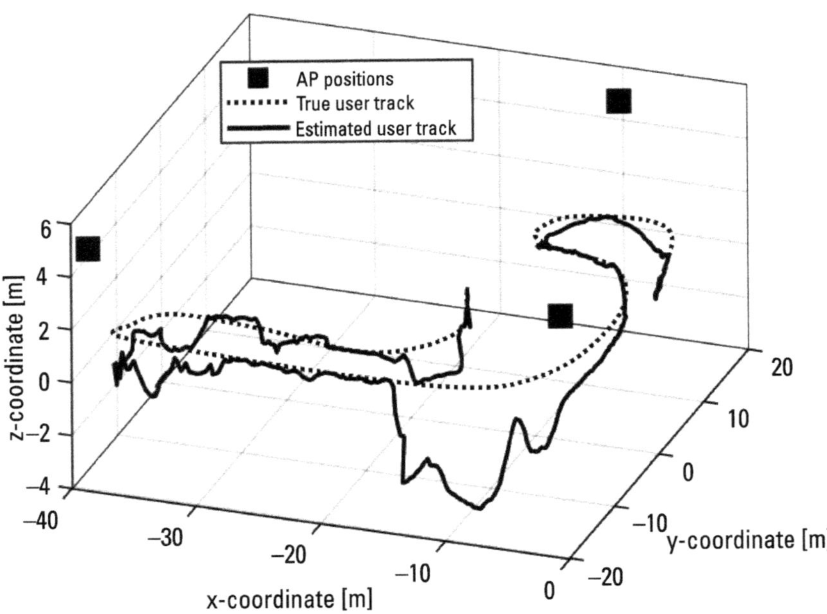

Figure 4.21 Three-dimensional illustration of the estimated user track using the EKF for FR2.

the CRB-based results, the poor estimation accuracy of user elevation is due to the bad geometry of the AP positions.

In the end, in Figure 4.22 we show the cumulative distribution of the EKF-based positioning error for the FR1 and FR2 setups while considering approaches with and without outlier detection. As already evident from the above visualizations of the estimated user tracks, the FR2 setup is shown to outperform the FR1 setup. Also, the positioning performances without the outlier detection is shown in the figure. Contrary to the previous results, the FR1 setup is able to perform slightly better without the outlier detection, but the overall performance is nevertheless unsuitable for modern indoor positioning applications.

According to the results and illustrations presented, it is clear that the management of a multipath channel, and the related measurement errors, is crucial for implementing an accurate indoor positioning system. Whereas CRB-based results can be used for theoretical evaluations as well as for tentative algorithm and system design, they are not able to appropriately capture the characteristics of the challenging indoor positioning environment.

An example MATLAB code is provided for simulating EKF-based positioning using only angle measurements, only range measurements, or both range and angle measurements for a chosen set of base stations. The example is built on a specific user track, where range and angle measurements are

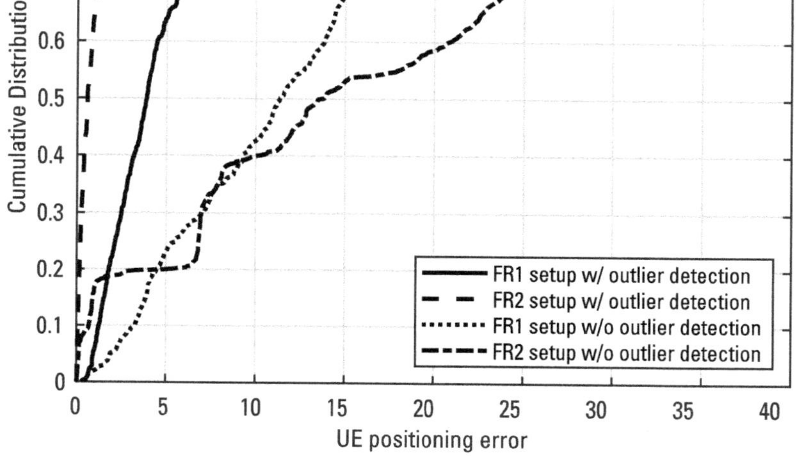

Figure 4.22 Cumulative distribution of positioning error for the FR1 and FR2 setup with and without outlier detection.

obtained assuming CRB-level estimation accuracy according to selected system parameters. However, the code can be modified to also cover other tracks and other measurement accuracies. However, to maintain reasonable positioning accuracy, modifying the simulation parameters might require also updating the EKF parameters.

4.6 Summary

This chapter started by looking at ways to improve accuracy and reliability of indoor navigation by using and sharing additional information. Map matching provides constraints from the environment; for example rooms, walls, and corridors indoors for improved position estimation. Particle filtering can use a wall crossing constraint as a measurement likelihood function; however, that comes with the price of higher computational complexity as well as the requirement for detailed map information. Computational efficiency is obtained when the floor plan and wall crossing constraints are replaced with graph-based constraints, turning the floor plan to a Voronoi graph, or a grid representation. When detailed maps are not available, they must be formed ad hoc. SLAM is an infrastructure-free technique, meaning that the user is able to simultaneously form a map of the unknown environment and locate themselves on it. Although SLAM development heavily concentrates on visual measurements, magnetometers or radio signals can be used when feasible, and when the setup is not constrained with requirements of being lightweight or computationally efficient, a combination of multiple sensors provides improved accuracy. Robustness of the obtained location and map are ensured by loop closure. Loop closure identifies revisits of locations saved in a database and corrects drift in the solution. At its most reduced, mapping in SLAM can be implemented based on learning motion patterns from IMU measurements when the user returns to a previously visited location. In cooperative navigation, each individual user estimates their navigation state and its uncertainty, as well as range and its uncertainty to others cooperating, communicates the results, and corrects and updates the solution accordingly. Cooperative navigation is especially powerful if some of the users have access to absolute solution, but it is able to improve the relative one, too.

Many applications, such as monitoring, surveillance, augmented reality, and sports need position information about others besides themselves and thereby tracking was discussed in this chapter. As with all the other topics discussed in this chapter, the development of tracking methods seems to be slowly transferring from Kalman filter based methods to deep learning. We ended the chapter and the book by providing an example of a complete radio-based navigation system that perfectly pulls together several themes of the book, with codes for the reader to get started with their own implementation serving as a basis for further development.

References

[1] Gu, F., X. Hu, M. Ramezani, D. Acharya, K. Khoshelham, S. Valaee, and J. Shang, "Indoor Localization Improved by Spatial Context–A Survey," *ACM Computing Surveys*, Vol. 52, No. 3, July 2019.

[2] Thorn, G., *Extracting Indoor Map Data from Public Escape Plans on Mobile Devices*, Master's Thesis, University of Munster, Germany, March 2013.

[3] Abdulrahim, K., C. Hide, T. Moore, and C. Hill, "Integrating Low Cost IMU with Building Heading in Indoor Pedestrian Navigation," *Journal of Global Positioning System*, Vol. 10, No. 1, 2011, pp. 30–38.

[4] Perttula, A., H. Leppakoski, M. Kirkko-Jaakkola, P. Davidson, J. Collin, and J. Takala, "Distributed Indoor Positioning System with Inertial Measurements and Map Matching," *IEEE Transactions on Instrumentation and Measurement*, Vol. 63, No. 11, 2014, pp. 2682–2695.

[5] Woodman, O., and R. Harle, "Pedestrian Localisation for Indoor Environments," in *Proceedings of the 10th International Conference on Ubiquitous Computing*, Seoul, Korea, 2008, pp. 114–123.

[6] Evennou, F., F. Marx, and E. Novakov, "Map-Aided Indoor Mobile Positioning System Using Particle Filter," in *Proceedings of the IEEE Wireless Communications and Networking Conference*, New Orleans, LA, March 2005, pp. 2490–2494.

[7] Liao, L., D. Fox, J. Hightower, H. Kautz, and D. Schulz, "Voronoi Tracking: Location Estimation Using Sparse and Noisy Sensor Data," in *Proceedings of the 2003 IEEE/RSJ International Conference on Intelligent Robots and Systems*, Las Vegas, NV, October 2003, pp. 723–728.

[8] Ferris, B., D. Hahnel, and D. Fox, "Gaussian Processes for Signal Strength-Based Location Estimation," in *Robotics: Science and Systems*, August 2006.

[9] Fox, D., W. Burgard, and S. Thrun, "Markov Localization for Mobile Robots in Dynamic Environments, *Journal of Artificial Intelligence Research*, Vol. 11, November 1999, pp. 391–427.

[10] Li, X., C. Claramunt, and C. Ray, "A Grid Graph-Based Model for the Analysis of 2D Indoor Spaces," *Computers, Environment and Urban Systems*, Vol. 34, No. 6, November 2010, pp. 532–540.

[11] Shang, J., X. Hu, W. Cheng, and H. Fan, "GridiLoc: A Backtracking Grid Filter for Fusing the Grid Model with PDR Using Smartphone Sensors," *Sensors*, Vol. 16, No. 12, 2016.

[12] Durrant-Whyte, H., and T. Bailey, "Simultaneous Localisation and Mapping (SLAM): Part I the Essential Algorithms," *IEEE Robotics and Automation Magazine*, Vol. 13, No. 2, 2006, pp. 99–110.

[13] Younes, G., D. Asmar, E. Shammas, and J. Zelek, "Keyframe-Based Monocular SLAM: Design, Survey, and Future Directions," *Robotics and Autonomous Systems*, Vol. 98, 2017, pp. 67–88.

[14] Mur-Artal and Tardos, ORB-SLAM2 Code, https://paperswithcode.com/method/orb-slam2.

[15] Ranftl, R., V. Vineet, Q. Chen, and V. Koltun, "Dense Monocular Depth Estimation in Complex Dynamic Scenes," in *Proceedings of the IEEE Conference on Computer Vision and Pattern Recognition*, 2016, pp. 4058–4066.

[16] Engel, J., and D. Cremers, "LSD-SLAM: Large-Scale Direct Monocular SLAM," https://vision.in.tum.de/research/vslam/lsdslam.

[17] Engel, J., V. Koltun, and D. Cremers, "DSO: Direct Sparse Odometry," https://vision.in.tum.de/research/vslam/dso/.

[18] Tsintotas, K., L. Bampis, and A. Gasteratos, "The Revisiting Problem in Simultaneous Localization and Mapping: A Survey on Visual Loop Closure Detection," *IEEE Transactions on Intelligent Transportation Systems*, Vol. 23, No. 11, 2022.

[19] A. J. Davison, I. D. Reid, N. D. Molton, and O. Stasse, "MonoSLAM: Real-Time Single Camera SLAM," *IEEE Transactions on Pattern Analysis and Machine Intelligence*, Vol. 29, No. 6, 2007, pp. 1052–1067.

[20] Botterill, T., S. Mills, and R. Green, "Correcting Scale Drift by Object Recognition in Single-Camera SLAM," *IEEE Transactions on Cybernetics*, Vol. 43, December 2013, pp. 1767–1780.

[21] Civera, J., A. J. Davison, and J. M. Martinez Montiel, "Inverse Depth Parametrization for Monocular SLAM," *IEEE Transactions on Robotics*, Vol. 24, No. 5, 2008, pp. 932–945.

[22] Campos, C., R. Elvira, J. J. Gomez Rodriguez, J. M. M. Montiel, and J. D. Tardos, "ORB-SLAM3: An Accurate Open-Source Library for Visual, Visual–Inertial, and Multimap SLAM," *IEEE Transactions on Robotics*, Vol. 37, No. 6, 2021, pp. 1874–1890.

[23] Favorskaya, M. N., "Deep Learning for Visual SLAM: The State-of-the-Art and Future Trends," *Electronics*, Vol. 12, No. 9, 2023.

[24] Zhu, R., M. Yang, W. Liu, R. Song, B. Yan, and Z. Xiao, "DeepAVO: Efficient Pose Refining with Feature Distilling for Deep Visual Odometry," *Neurocomputing*, Vol. 467, 2022, pp. 22–35.

[25] Almalioglu, Y., M. Turan, M. R. U. Saputra, P. P. B. de Gusmao, A. Markham, and N. Trigoni, "SelfVIO: Self-Supervised Deep Monocular Visual–Inertial Odometry and Depth Estimation," *Neural Networks*, 150:119–136, 2022.

[26] Li, G., L. Yu, and S. Fei, "A Deep-Learning Real-Time Visual Slam System Based on Multi-Task Feature Extraction Network and Self-Supervised Feature Points," *Measurement*, Vol. 168, 2021.

[27] Li, R., S. Wang, and D. Gu, "DeepSLAM: A robust Monocular SLAM System with Unsupervised Deep Learning," *IEEE Transactions on Industrial Electronics*, Vol. 68, No. 4, 2021, pp. 3577–3587.

[28] Naveed, K., M. L. Anjum, W. Hussain, et al., "Deep Introspective SLAM: Deep Reinforcement Learning Based Approach to Avoid Tracking Failure in Visual SLAM," *Autonomous Robots*, Vol. 46, 2022, pp. 705–724.

[29] Xiu, H., Y. Liang, H. Zeng, Q. Li, H. Liu, B. Fan, and C. Li, "Robust Self-Supervised Monocular Visual Odometry Based on Prediction-Update Pose Estimation Network," *Engineering Applications of Artificial Intelligence*, Vol. 116, 2022, p. 105481.

[30] Wahlstrom, N., M. Kok, T. B. Schon, and F. Gustafsson, "Modeling Magnetic Fields Using Gaussian Processes," in *Proceedings of the IEEE International Conference on Acoustics, Speech and Signal Processing*, Vancouver, BC, Canada, May 2013, pp. 3522–3526.

[31] Kok, M., and A. Solin, "Scalable Magnetic Field SLAM in 3D Using Gaussian Process Maps," in *Proceedings of the 21st International Conference on Information Fusion (FUSION)*, July 2018, pp. 1353–1360.

[32] Skog, I., G. Hendeby, and F. Trulsson, "Magnetic-Field Based Odometry—An Optical Flow Inspired Approach," in *Proceedings of the International Conference on Indoor Positioning and Indoor Navigation (IPIN)*, Lloretde Mar, Spain, November 2021.

[33] Ferris, B., D. Fox, and N. D. Lawrence, "WiFi-SLAM Using Gaussian Process Latent Variable Models," in *Proceedings of the 20th International Joint Conference on Artificial Intelligence (IJCAI)*, Hyderabad, India, January 2007, pp. 2480–2485.

[34] Gutmann, J.-S., E. Eade, P. Fong, and M. E. Munich, "Vector Field SLAM—Localization by Learning the Spatial Variation of Continuous Signals," *IEEE Transactions on Robotics*, Vol. 28, No. 3, 2012, pp. 650–667.

[35] Huang, J., D. Millman, M. Quigley, D. Stavens, S. Thrun, and A. Aggarwal, "Efficient, Generalized Indoor WiFi GraphSLAM," in *IEEE International Conference on Robotics and Automation*, Shanghai, China, May 2011, pp. 1038–1043.

[36] Bruno, L., and P. Robertson, "WiSLAM: Improving FootSLAM with WiFi," in *International Conference on Indoor Positioning and Indoor Navigation*, Guimaraes, Portugal, September 2011.

[37] Angermann, M., and P. Robertson, "FootSLAM: Pedestrian Simultaneous Localization and Mapping without Exteroceptive Sensors—Hitchhiking on Human Perception and Cognition," *Proceedings of the IEEE*, Vol. 100, April 2012, pp. 1840–1848.

[38] Wymeersch, H., J. Lien, and M. Z. Win, "Cooperative Localization in Wireless Networks," *Proceedings of the IEEE*, Vol. 97, No. 2, 2009, pp. 427–450.

[39] da Costa, A. L., and G. Bittencourt, "Cooperative Mobile Robots: A Cognitive Multi-Agent Approach," *IFAC Proceedings Volumes*, Vol. 39, No. 20, 2006, pp. 77–82.

[40] Li, C., L. Shi, N. Moayeri, and J. Benson, "A Performance Comparison of Wi-Fi RTT and UWB for RF Ranging," National Institute of Standards and Technology (NIST), 2020.

[41] Morrison, A. J., L. Ruotsalainen, M. Makela, J. Rantanen, and N. Sokolova, "Combining Visual, Pedestrian, and Collaborative Navigation Techniques for Team Based Infrastructure Free Indoor Navigation," *Proceedings of the 32nd International Technical Meeting of the Satellite Division of The Institute of Navigation (ION GNSS+ 2019)*, 2019, pp. 2692–2701.

[42] Makela, M., M. Kirkko-Jaakkola, J. Rantanen, and L. Ruotsalainen, "Proof of Concept Tests on Cooperative Tactical Pedestrian Indoor Navigation," *21st International Conference on Information Fusion (FUSION)*, 2018, pp. 1369–1376.

[43] Lucas, B., and T. Kanade, "An Iterative Image Registration Technique with an Application to Stereo Vision," *JCAI'81: 7th International Joint Conference on Artificial Intelligence*, Vancouver, Canada, August 1981, pp. 674–679.

[44] Kristan, M., A. Leonardis, J. Matas, et al., "The Eighth Visual Object Tracking Vot2020 Challenge Results," in *Computer Vision–ECCV 2020 Workshops, Proceedings, Part V 16*, Glasgow, U.K., August 23–28, 2020, pp. 547–601.

[45] VOT. Visual Object Tracking (VOT), https://www.votchallenge.net/index.html.

[46] Chen, F., X. Wang, Y. Zhao, S. Lv, and X. Niu, "Visual Object Tracking: A Survey," *Computer Vision and Image Understanding*, Vol. 222, 2022, pp. 103508.

[47] Smeulders, A. W. M., D. M. Chu, R. Cucchiara, S. Calderara, A. Dehghan, and M. Shah, "Visual Tracking: An Experimental Survey," *IEEE Transactions on Pattern Analysis and Machine Intelligence*, Vol. 36, No. 7, 2014, pp. 1442–1468.

[48] Gavves, E., and D. Gupta, "Long-Term Deep Object Tracking," in *Advanced Methods and Deep Learning in Computer Vision* (E.R. Davies and M. A. Turk, eds.), London, U.K.: Academic Press, 2022, pp. 337–371.

[49] Wojke, N., A. Bewley, and D. Paulus, "Simple Online and Real-Time Tracking with a Deep Association Metric," in *2017 IEEE International Conference on Image Processing (ICIP)*, 2017, pp. 3645–3649.

[50] Shuai, B., A. Berneshawi, X. Li, D. Modolo, and J. Tighe. Siammot, "Siammot: Siamese Multi-Object Tracking," in *Proceedings of the IEEE/CVF Conference on Computer Vision And Pattern Recognition*, 2021, pp. 12372–12382.

[51] Yan, B., H. Peng, J. Fu, D. Wang, and H. Lu, "Learning Spatio-Temporal Transformer for Visual Tracking," *Proceedings of the IEEE/CVF International Conference on Computer Vision (ICCV)*, 2021, pp. 10448–10457.

[52] Talvitie, J., M. Valkama, G. Destino, and H. Wymeersch, "Novel Algorithms for High-Accuracy Joint Position and Orientation Estimation in 5G MM wave Systems," in *2017 IEEE Globecom Workshops (GC Wkshps)*, 2017 pp. 1–7.

[53] MacCartney, G. R., and T. S. Rappaport, "Rural Microcell Path Loss Models for Millimeter Wave Wireless Communications," *IEEE Journal on Selected Areas in Communications*, Vol. 35, No. 7, 2017, pp. 1663–1677.

[54] Oda, Y., K. Tsunekawa, and M. Hata, "Advanced LOS Path-Loss Model in Microcellular Mobile Communications," *IEEE Transactions on Vehicular Technology*, Vol. 49, No. 6, 2000, pp. 2121–2125.

[55] Samimi, M. K., T. S. Rappaport, and G. R. MacCartney, "Probabilistic Omnidirectional Path Loss Models for Millimeter-Wave Outdoor Communications," *IEEE Wireless Communications Letters*, Vol. 4, No. 4, 2015, pp. 357–360.

[56] Zhang, Y., L. Pang, G. Ren, F. Gong, X. Liang, J. Dou, and J. Li, "3-D MIMO Parametric Stochastic Channel Model for Urban Macrocell Scenario," *IEEE Transactions on Wireless Communications*, Vol. 16, No. 7, 2017, pp. 246–4260.

[57] Zwick, T., C. Fischer, D. Didascalou, and W. Wiesbeck, "A Stochastic Spatial Channel Model Based on Wave-Propagation Modeling," *IEEE Journal on Selected Areas in Communications*, Vol. 18, No. 1, 2000, pp. 6–15.

[58] European Telecommunications Standards Institute (ETSI), *TR 138 901,5G; Study on Channel Model for Frequencies from 0.5 to 100 GHz*, 2017.

[59] Nurmela, V., A. Karttunen, A. Roivainen, et al., "Deliverable D1. 4 METIS Channel Models," in *Proceedings of the Mobile Wireless Communications Enablers Information Society (METIS)*, 2015, p. 1.

[60] Chen, Y., J. Palacios, N. Gonzalez-Prelcic, T. Shimizu, and H. Lu, "Joint Initial Access and Localization in Millimeter Wave Vehicular Networks: A Hybrid Model/Data Driven Approach," 2022, *arXivpreprint arXiv:2204.01510.*

[61] Ge, Y., O. Kaltiokallio, H. Kim, et al., "A Computationally Efficient EK-PMBM Filter for Bistatic mmWave Radio SLAM," *IEEE Journal on Selected Areas in Communications*, Vol. 40, No. 7, 2022, pp. 2179–2192.

[62] Stremler, F. G., *Introduction to Communication Systems*, Addison-Wesley, 1990.

[63] Kirubarajan, T., Y. Bar-Shalom, and X. Rong Li, *Estimation with Applications to Tracking and Navigation: Theory, Algorithms and Software*, New York: Wiley Interscience, 2001.

[64] Tichavsky, P., C. H. Muravchik, and A. Nehorai, "Posterior Cramer-Rao Bounds for Discrete-Time Nonlinear Filtering," *IEEE Transactions on Signal Processing*, Vol. 46, No. 5, 1998, pp. 1386–1396.

[65] Sand, S., A. Dammann, and C. Mensing, *Positioning in Wireless Communications Systems*, Chichester, U.K.: Wiley, 2014.

[66] Li, B., S. Wang, J. Zhang, X. Cao, and C. Zhao, "Fast Randomized-Music for mm-Wave Massive MIMO Radars," *IEEE Transactions on Vehicular Technology*, Vol. 70, No. 2, 2021, pp. 1952–1956.

[67] Zheng, Q., L. Luo, H. Song, G. Sheng, and X. Jiang, "A RSSI-AOA-based UHF Partial Discharge Localization Method Using Music Algorithm," *IEEE Transactions on Instrumentation and Measurement*, Vol. 70, 2021, pp. 1–9.

[68] Chuang, S.-F., W.-R. Wu, and Y.-T. Liu, "High-Resolution AOA Estimation for Hybrid Antenna Arrays," *IEEE Transactions on Antennas and Propagation*, Vol. 63, No. 7, 2015, pp. 2955–2968.

[69] Talvitie, J., T. Levanen, M. Koivisto, K. Pajukoski, M. Renfors, and M. Valkama, "Positioning of High-Speed Trains Using 5G New Radio Synchronization Signals," in *2018 IEEE Wireless Communications and Networking Conference (WCNC)*, 2018, pp. 1–6.

[70] Ding H., and K. G. Shin, "Accurate Angular Inference for 802.11ad Devices Using Beam-Specific Measurements," *IEEE Transactions on Mobile Computing*, Vol. 21, No. 3, 2022, pp. 822–834.

[71] Talvitie, J., T. Levanen, M. Koivisto, T. Ihalainen, K. Pajukoski, M. Renfors, and M. Valkama, "Positioning and Location-Based Beamforming for High Speed Trains in 5G NR Networks," in *2018 IEEE Globecom Workshops (GC Wkshps)*, 2018, pp. 1–7.

List of Abbreviations

E_c/N_0	Energy per chip over the noise spectral density
3GPP	3rd Generation Partnership Project
A-GNSS	Assisted GNSS
ADEV	Allan deviation
AOA	Angle of arrival
AOD	Angle of departure
AP	Access point
ARW	Angle random walk
BLE	Bluetooth Low Energy
BPM	Burst position modulation
BPSK	Binary phase-shift keying
BS	Base station
CDMA	Code division multiple access
CPICH	Common pilot channel
CRB	Cramér–Rao bound
CRS	Cell-specific reference signal
CTE	Constant tone extension
DL	Downlink
DL-AOD	Downlink angle of departure
DL-TDOA	Downlink time difference of arrival
DPCCH	Dedicated physical control channel
DPCH	Downlink dedicated physical channel
DR	Dead reckoning
Ds-TWR	Double-sided two-way ranging
EKF	Extended Kalman filter

eNB	Evolved Node B
ENU	East–North–Up
EOTD	Enhanced observed time difference
ERDEV	Enhanced ranging-capable device
ETSI	European Telecommunications Standards Institute
EXIF	Exchangeable image file
FDD	Frequency division duplex
FFT	Fast Fourier transform
FGO	Factor graph optimization
FHSS	Frequency hopping spread spectrum
FIM	Fisher Information Matrix
FOG	Fiber-optic gyroscope
FR1	Frequency Range 1
FR2	Frequency Range 2
FTM	Fine time measurement
GFSK	Gaussian frequency shift keying
GLRT	Generalized likelihood ratio test
GMSK	Gaussian minimum shift keying
gNB	Next Generation Node B
GNSS	Global Navigation Satellite System
GPRS	General Packet Radio Service
GSM	Global System for Mobile Communications
HRP	High-rate pulse repetition frequency
HSGNSS	High sensitivity GNSS
IEEE	Institute of Electrical and Electronics Engineers
IMU	Inertial measurement unit
ISM	Industrial, scientific, and medical
ITU-R	International Telecommunication Union Radiocommunication Sector
KF	Kalman filter
LiDAR	Light detection and ranging
LMU	Location measurement unit
LOS	Line of sight
LRP	Low-rate pulse repetition frequency
LS	Least squares
LTE	Long Term Evolution
MAC	Medium Access Control
MEMS	Microelectromechanical system
MIMO	Multiple-input and multiple-output
ML	Machine learning
MS	Mobile station
Multi-RTT	Multicell round-trip time

MUSIC	Multiple signal classification
NGP	Next Generation Positioning
NLOS	Non-line-of-sight
NR	New Radio
OFDM	Orthogonal frequency-division multiplexing
OFDMA	Orthogonal frequency-division multiple access
OSI	Open Systems Interconnection
OTD	Observed time difference
OTDOA	Observed time difference of arrival
PCB	Printed circuit board
PDR	Pedestrian dead reckoning
PEB	Position error bound
PF	Particle filter
PHY	Physical
PRS	Positioning reference signal
PSK	Phase shift keying
QAM	Quadrature amplitude modulation
RDEV	Ranging-capable device
RF	Radio frequency
RGB-D	Red green blue-distance
RLG	Ring laser gyroscope
RP	Reference points
RSCP	Received signal code power
RSRP	Reference signal received power
RSRQ	Reference signal received quality
RSSI	Received signal strength indicator
RSTD	Reference signal time difference
RTD	Real time difference
RTT	Round-trip time
SFN	System frame number
SHOE	Stance hypothesis optimal detection
SIFT	Scale-Invariant Feature Transform
SIG	Special interest group
SLAM	Simultaneous localization and mapping
SNR	Signal-to-noise ratio
SoOp	Signals-of-opportunity
SRS	Sounding reference signal
SS	Synchronization signals
Ss-TWR	Single-sided two-way ranging
SSB	Synchronization signal block
SSD	Sum of squared differences
STA	Station

TA	Timing advance
TDD	Time division duplex
TDMA	Time division multiple access
TDOA	Time difference of arrival
TOA	Time of arrival
TOF	Time of flight
TRL	Technology readiness level
TWR	Two-way ranging
U-TDOA	Uplink-time difference of arrival
UE	User equipment
UL	Uplink
UL-AOA	Uplink angle of arrival
UL-RTOA	Uplink relative time of arrival
UL-TDOA	Uplink time difference of arrival
ULA	Uniform linear array
UMTS	Universal Mobile Telecommunications System
UTOA	Uplink time of arrival
UWB	Ultrawideband
VRW	Velocity random walk
WLAN	Wireless local area network
ZUPT	Zero-velocity update

List of Symbols

(x, y)	Location of pixel
(u, v)	Pixel velocity
(x_0, y_0)	Principal point (camera's intrinsic parameter)
(x_c, y_c)	Corrected image points
(x_d, y_d)	Normalized distorted image point
$\alpha_{C,j}$	A priori parameter
Δd	Propagation distance difference between two consecutive antenna elements
Δt	Interval between two consecutive time epochs
δt	Change in time
δx_k	x-coordinate offset
δy_k	y-coordinate offset
δz	Error in object's depth
δz_k	z-coordinate offset
$\delta(\cdot)$	Unit impulse function
$\delta\boldsymbol{\psi}$	Attitude error
$\delta\boldsymbol{p}$	Position error
$\delta\boldsymbol{v}$	Velocity error
Δ_f	Subcarrier spacing
$\dot{\mathrm{R}}_A^B$	Time derivative of R_A^B
γ	SNR of the received signal, or threshold value (dead reckoning)
$\hat{\tau}$	Estimated propagation time/delay
$\hat{T}_{propagDS}$	Estimated propagation delay using Ds-TWR
$\hat{T}_{propagSS}$	Estimated propagation delay using Ss-TWR
λ	Signal wavelength

Λ	Geographical latitude
$H(\boldsymbol{x})$	Jacobian matrix of measurement model with respect to position $\boldsymbol{x}$
$H[k]$	Jacobian matrix (radio-based positioning)
$H_{AZ\text{-}AOA}[k]$	Jacobian matrix for azimuth AOA part (radio-based positioning)
$H_{EL\text{-}AOA}[k]$	Jacobian matrix for elevation AOA part (radio-based positioning)
$H_f(f)$	Antenna-element-wise frequency response
$H_{OFDM}(n)$	Subcarrier-wise frequency response
$H_r(\tau)$	Channel impulse response
$H_{TOA}[k]$	Jacobian matrix for TOA part (radio-based positioning)
$J(\boldsymbol{x})$	Fisher information matrix of parameter $\boldsymbol{x}$
R	Rotation matrix
R_B^L	Rotation matrix (from B-frame to L-frame)
Ω_E	Earth rotation rate with respect to the I-frame
ω_i	ith element of angular rate vector
Ω_n	Set of averaging indices (dead reckoning)
ϕ	Drift angle (dead reckoning)
$p\,(h)$	Air pressure
$\Psi(\cdot)$	Antenna pattern function
ψ	Heading (or yaw)
ψ	Phase parameter related to the received signal
σ_a	Accelerometer noise variance
σ_D	Noise density
σ_w^2	Noise variance
σ_v^2	Variance of user acceleration (radio-based positioning)
$[\boldsymbol{\omega}^B]_\times$	Cross product matrix
τ	Propagation time/delay, or averaging time (inertial sensors)
τ_k	Propagation delay of kth path (radio channel)
d	Direction
E	Essential matrix
F	Fundamental matrix (for epipolar geometry)
H	Kernel (in convolution)
H_k	Observation matrix (cooperative navigation)
K	Camera calibration matrix
K_i	Camera calibration matrix at the time of taking the ith image
l2	Epipolar line
l_i	ith landmark
Pk	Camera model matrix at epoch k
P	Camera model matrix
t	Translation vector
$T_{ReplyA/B}$	Delay measurement in UWB ranging
$T_{RoundTripA/B}$	RTT measurement in UWB ranging

$\mathbf{u}_k$	Control vector (probabilistic SLAM)
$\mathbf{v}$	Vanishing point
$\mathbf{X}$	3D object point
$\mathbf{X}_c$	Camera-centered 3D point
$\mathbf{x}_k$	User state vector (probabilistic SLAM)
$\mathbf{x}_i$	ith image point
$\mathbf{z}_{ki}$	Observations of landmarks' locations and orientations
b	Baseline
D	Lens aperture
fx	Focal length in x
fy	Focal length in y
f	Focal length
$I(x,y)$	Intensity value of the pixel at location x,y
$I(x,y,t)$	Intensity value of the pixel at location x,y and time t
I	Image
z	Object's depth
θ	Elevation angle
θ_i	Elevation angle observed at ith reference node
$\theta_{\text{AOA},k}$	Elevation AOA of kth path (radio channel)
$\theta_{\text{AOD},k}$	Elevation AOD of kth path (radio channel)
φ	Azimuth angle
φ_i	Azimuth angle observed at ith reference node
$\varphi_{\text{AOA},k}$	Azimuth AOA of kth path (radio channel)
$\varphi_{\text{AOD},k}$	Azimuth AOD of kth path (radio channel)
$\boldsymbol{\epsilon}_m$	Measurement noise in magnetometer
$\boldsymbol{\epsilon}_f$	Measurement errors in accelerometer
$\eta_{\text{AOA},k}$	3D AOA of kth path (radio channel)
$\eta_{\text{AOD},k}$	3D AOD of kth path (radio channel)
$\boldsymbol{\omega}_{BI}^{B}$	Angular rate vector in B-frame (rotation of B-frame with respect to I-frame)
$\boldsymbol{\omega}_{EI}$	Earth rotation rate
$\boldsymbol{\omega}_{LE}$	Transport rate
$\boldsymbol{\theta}$	Parameter vector
$\boldsymbol{a}$	Acceleration
$\boldsymbol{a}^B$	Acceleration (body frame)
$\boldsymbol{a}_{\text{RX}}(\cdot)$	Steering vector for receiver array
$\boldsymbol{a}_{\text{TX}}(\cdot)$	Steering vector for transmitter array
$\boldsymbol{b}_{\text{RX}}$	Receiver-side beamformer
$\boldsymbol{b}_{\text{TX}}$	Transmitter-side beamformer
f^B	Force measurement vector (body frame)
g^L	Gravitational acceleration (local frame)

$\boldsymbol{h}(\boldsymbol{x})$	Measurement model at position $\boldsymbol{x}$
$\boldsymbol{h}$	Hard iron anomaly
$\boldsymbol{m}^B$	Three-axis magnetometer measurement
$\boldsymbol{M}^L$	Earth magnetic field vector
$\boldsymbol{n}_{\mathrm{RX}}(n)$	Measurement noise (radio-based positioning)
$\boldsymbol{p}_{\mathrm{AN},m}$	Position of mth access node (radio-based positioning)
$\boldsymbol{p}_{\mathrm{u}}[k]$	3D user position at epoch k (radio-based positioning)
$\boldsymbol{p}_t$	Position at epoch t
$\boldsymbol{s}_{\mathrm{TX}}(n)$	Transmitted signal (radio-based positioning)
$\boldsymbol{v}_{\mathrm{u}}[k]$	3D user velocity at epoch k (radio-based positioning)
$\boldsymbol{v}_t$	Velocity at epoch t
$\boldsymbol{x}$	User position
$\boldsymbol{y}_{\mathrm{RX}}(n)$	Received signal vector (radio-based positioning)
wi_t	Particle weight at epoch t
ζ_m	Phase-shift applied to the mth antenna element
A	Amplitude parameter related to the received signal
B	Bandwidth
b	Bias (neural networks)
b_z	Barometer bias
C	Cost matrix (computer vision)
c	Speed of light
C_{RSSI_r}	Occurrence of $RSSI_r$ in database training set
$d(p, q)$	Euclidean distance between p and q (L_2-norm)
D	Fingerprinting database
d	Disparity (amount of pixel's horizontal motion)
d_i	Range between user and ith reference node
$d_{2\mathrm{D},i}$	2D distance between user and ith reference node
d_{ant}	Antenna separation distance
E	Graph edges (computer vision)
f_s or f_s	Sampling frequency
F_s	Sampling frequency
F_{dB}	Noise figure
G	Filtered image (after convolution with specific kernel)
g	Gravitational acceleration
G_R	Receiver antenna gain
G_T	Transmitter antenna gain
h	Altitude
h_0	Reference height
h_k	kth path coefficient (radio channel)
$h_{\mathrm{t}}(\tau)$	Effective scalar channel impulse response
I_t	Temporal gradient

I_x	Partial derivative with respect to x
K	Number of subcarriers or number of multipaths (radio channel)
k_i	Distortion values specific to the camera
M	Number of antenna elements
M_0	Molar mass of air
m_E	Magnetic field strength in East-direction at position $\boldsymbol{x}$
m_N	Magnetic field strength in North-direction at position $\boldsymbol{x}$
m_U	Magnetic field strength in Up-direction at position $\boldsymbol{x}$
N_{RX}	Number of antenna elements in receiver
N_{TX}	Number of antenna elements in transmitter
$N_{C,j}^{(i)}$	Number of times the ith particle has crossed the jth edge of cell C
p	Pitch
P_m	Power measurement from the mth sector
P_R	Received power
P_T	Transmit power
P_{n}	Thermal noise power (radio-based positioning)
p_{t_k}	Position at time epoch t_k (dead reckoning)
$r(n)$	Received signal sample
R	Universal gas constant
r	Roll or radial distance of corrected image point
r_d	Radial distance of normalized distorted image points
R_i	Fingerprint measurement
r_m	Received signal at the mth antenna element
$S(f)$	Spectrum of transmitted signal
S	Soft iron distortion or skew coefficient (camera's intrinsic parameter)
S_k	Transmitted symbol at kth subcarrier
$T(z_n)$	Logarithm of likelihood ratio (dead reckoning)
T_0	Temperature at reference height
T_c	5G NR time unit
t_i	ith time epoch, time stamp, or time instant
T_s	Sample interval or LTE time unit
T_{replyA}	Reply time measured at device A
T_{replyB}	Reply time measured at device B
T_{reply}	Measured reply time
T_{roundA}	Round-trip time measured at device A
T_{roundB}	Round-trip time measured at device B
T_{round}	Round-trip time
V	Graph vertices/nodes (computer vision)
w_i	ith input weight (neural networks)
w_m	Additive white Gaussian noise

$w_{is}(x_u, y_u)$	Weight of *i*th reference point (fingerprinting)
x_i	*i*th angle or velocity value (inertial sensors), or input value (neural networks)
x_u	User *x*-coordinate
x_m	*x*-coordinate of *m*th access node (radio-based positioning)
x_{ti}	*x*-coordinate of *i*th reference node
y	Neural network function)
y_i	*i*th measurement
y_k^{ω}	Measurement of angular rate at time epoch *k* (dead reckoning)
y_k^{a}	Measurement of specific force at time epoch *k* (dead reckoning)
y_u	User *y*-coordinate
y_m	*y*-coordinate of *m*th access node (radio-based positioning)
y_{ti}	*y*-coordinate of *i*th reference node
Z	Distance between camera and object
z_u	User *z*-coordinate
z_m	*z*-coordinate of *m*th access node (radio-based positioning)
z_{ti}	*z*-coordinate of *i*th reference node

About the Authors

Laura Ruotsalainen is a professor in computer science at the University of Helsinki, Finland. She leads a research group in spatiotemporal data analysis for sustainability science (SDA) that does research on estimation and machine learning methods using spatiotemporal data. She has a long research career in the navigation field, including GNSS and sensor fusion for urban and indoor environments, computer vision, and analysis of GNSS signal characteristics and GNSS interference mitigation. She is a member of the steering group of the Finnish Center for AI (FCAI). She received her master's degree from the Department of Computer Science, University of Helsinki, in 2003 and doctoral degree in 2013 from the Department of Pervasive Computing, Tampere University of Technology. Her doctoral research was partly done at the University of Calgary, Canada.

Martti Kirkko-Jaakkola is a research manager in the Department of Navigation and Positioning at the Finnish Geospatial Research Institute, National Land Survey of Finland. Since 2019, he also works for Nordic Inertial Oy, Finland. He received his MSc and DSc (Tech) degrees from Tampere University of Technology, Finland, in 2008 and 2013, respectively. He started his career in the field of positioning and navigation as a summer trainee in 2006, and has worked on various projects ranging from satellite positioning and timing to inertial navigation and sensor fusion. He has also served as an external project reviewer for the European GNSS Agency within the Horizon 2020 program. Dr. Kirkko-Jaakkola is an editorial board member for *GPS Solutions* and an associate editor for *IEEE Transactions on Instrumentation and Measurement.*

Jukka Talvitie is currently a university lecturer at the Unit of Electrical Engineering in Tampere University, Finland, working in the field of wireless communications, radio positioning, and radio-based sensing, focusing on 5G NR and future wireless networks. He has more than 80 international peer-reviewed scientific publications, including journals, conference proceedings, and book chapters. In addition, he has contributed to more than 20 patents or patent applications and has acquired extensive experience in working in industry, such as in Nokia, HERE technologies, and Renesas. He has supervised more than 25 BSc/MSc students, and more than 5 PhD students. He is currently leading research projects with positioning emphasis funded by the European Space Agency (ESA) and Academy of Finland. His research interests include signal processing for wireless communications, network-based positioning methods, radio-based sensing and mapping, simultaneous localization and mapping, device tracking and filtering methods, and machine-learning methods for wireless communications, positioning, and sensing.

Index

Artech House
GNSS Technology and Applications Library

Elliott Kaplan and Christopher Hegarty, Series Editors

A-GPS: Assisted GPS, GNSS, and SBAS, Frank van Diggelen

All Source Positioning, Navigation, and Timing, Rongsheng (Ken) Li

Applied Satellite Navigation Using GPS, GALILEO, and Augmentation Systems, Ramjee Prasad and Marina Ruggieri

Digital Terrain Modeling: Acquisition, Manipulation, and Applications, Naser El-Sheimy, Caterina Valeo, and Ayman Habib

Geographical Information Systems Demystified, Stephen R. Galati

GNSS Applications and Methods, Scott Gleason and Demoz Gebre-Egziabher

GNSS Interference Threats and Countermeasures, Fabio Dovis, editor

GNSS Markets and Applications, Len Jacobson

GNSS Receivers for Weak Signals, Nesreen I. Ziedan

GNSS for Vehicle Control, David M. Bevly and Stewart Cobb

GPS/GNSS Antennas, B. Rama Rao, W. Kunysz, R. Fante, and K. McDonald

Implementing e-Navigation, John Erik Hagen

Inertial Navigation Systems Analysis, Kenneth Britting

Introduction to GPS: The Global Positioning System, Second Edition, Ahmed El-Rabbany

Location-Based Services in Cellular Networks: From GSM to 5G NR, Adrián Caldaldo García, Stefan Maier, and Abhay Phillips

MEMS-Based Integrated Navigation, Priyanka Aggarwal, Zainab Syed, Aboelmagd Noureldin, and Naser El-Sheimy

Navigation Signal Processing for GNSS Software Receivers, Thomas Pany

The Present and Future of Indoor Navigation, Laura Ruotsalainen, Martti Kirkko-Jaakkola, and Jukka Talvitie

Principles of GNSS, Inertial, and Multisensor Integrated Navigation Systems, Second Edition, Paul D. Groves

Radionavigation Systems, Borje Forssell

RF Positioning: Fundamentals, Applications, and Tools, Rafael Saraiva Campos, and Lisandro Lovisolo

Server-Side GPS and Assisted-GPS in Java™, Neil Harper

Spread Spectrum Systems for GNSS and Wireless Communications, Jack K. Holmes

Understanding GPS/GNSS: Principles and Applications, Third Edition, Elliott Kaplan and Christopher Hegarty, editors

Ubiquitous Positioning, Robin Mannings

Wireless Positioning Technologies and Applications, Second Edition, Alan Bensky

For further information on these and other Artech House titles, including previously considered out-of-print books now available through our In-Print-Forever® (IPF®) program, contact:

Artech House Publishers
685 Canton Street
Norwood, MA 02062
Phone: 781-769-9750
Fax: 781-769-6334
e-mail: artech@artechhouse.com

Artech House Books
16 Sussex Street
London SW1V 4RW UK
Phone: +44 (0)20 7596 8750
Fax: +44 (0)20 7630 0166
e-mail:
artech-uk@artechhouse.com

Find us on the World Wide Web at: www.artechhouse.com